W9-AOE-043

WATCHES TELL MORE THAN TIME

Product Design, Information, and the Quest for Elegance

DEL COATES

McGRAW-HILL

New York Chicago San Francisco Lisbon London Madrid Mexico City Milan New Delhi San Juan Seoul Singapore Sydney Toronto

McGraw-Hill

A Division of The McGraw-Hill Companies

1 2 3 4 5 6 7 8 9 0 DOC/DOC 0 9 8 7 6 5 4 3 2

ISBN 0-07-136243-6

This book was set in Optima by North Market Street Graphics.

Printed and bound by R. R. Donnelley & Sons Company.

This book is printed on recycled, acid-free paper containing a minimum of 50% recycled, de-inked fiber.

To Betsy,
Leslie,
Tracy,
and Jan

Contents

Introduction

OBVIOUSLY THERE IS A NEW ART, existent in machine-made mass products: industrial design. It is not an esoteric and precious manifestation but a practical expression embodied in utilitarian forms increasingly familiar in the daily life of the average person. Everywhere it condemns the standards of taste by which we formerly chose our furnishings and our "ornaments" and foreshadows a new universal style. But the manifestation is so new that there has been hardly an effort to evaluate it, to trace its origins, or to fix its boundaries. . . . we have seen the beginnings of a new and truly creative art and the foundation of a new national and international culture which is a product of the mechanized world and rich in spiritual resources [pp. vii–xi].

—Sheldon and Martha Cheney

Art and the Machine:
An Account of Industrial Design in 20th Century America

Arthur Pulos notes, in *American Design Ethic: A History of Industrial Design,* that U.S. Commissioner of Patents Edward B. Moore coined the term *industrial design* in 1913. By the 1920s, when interest in design of anything and everything was high, *industrial designer* served aptly to distinguish those who designed industrially produced products from graphic designers, interior designers, fashion designers, and architects. The word *industrial* also had considerable cachet during the roaring twenties, which roared in large measure because rampant industrialization and technological progress were fulfilling the Utopian promises of the industrial revolution. Thus the 1930s were an especially auspicious decade for the young field to take root and flourish. The decade turned out to be the most fervently creative and innovative decade yet—technically, socially, and artistically. Industrialization flowed from and symbolized progress in science, technology, and—with the advent of industrial design—spatial art.

As matters turned out, the Cheneys' expectations proved overly optimistic or, at least, premature. Industrial designers have remained pretty much the invisible members of the product design and development team throughout the intervening period—until lately. With rising interest in product aesthetics—in boardrooms and living rooms—some observers are already calling this another golden age of industrial design. Certainly, there has never been more interest in product design since the 1930s. The field of industrial design has never seemed more valuable, accomplished, or promising. Having emerged just as the industrial revolution gained its stride, the computers and technologies of the information age now shape its destiny.

My lifelong fascination with product design—in particular, with product aesthetics—began as long ago as I can remember, when I first fell under a beautiful car's spell. I began my quest in the psychological literature for what philosopher George Santayana called the "sense of beauty" before I was old enough to drive. At twenty, I published my first article, confidently titled "Why Some Cars Are Beautiful," the same year I realized my dream of becoming a car designer.

The other stream shaping this book had surfaced in 1949, with publication of Shannon's and Weaver's *The Mathematical Theory of Communication.* The two streams—information science and what came to be known as *psychological aesthetics*—converged for me in 1965. As a designer with Ford Motor Company's Advanced Vehicle Concepts Department, I saw a presentation of animated computer graphics by MIT doctoral candidate Ivan Sutherland, who later came to be revered as the "grandfather of computer graphics." He projected images of simple geometric solids rotating on the screen. However, I saw the future of automotive design—photographically realistic images of virtual cars rotating on virtual turntables.

Computer-aided design (CAD) and computer-aided engineering (CAE) had already established their future stakes in the automobile industry. Design constraints were accumulating at such a rampant rate—due chiefly to the advent of government safety regulations—that costs for design and development, in terms of complexity, time, and cash, would compound exponentially if engineers continued to design cars as they had in the past with T-squares, triangles, and pencils. Matters would become even more hectic a decade later in the wake of the energy crises of the 1970s as federal fuel economy standards came into being. CAD/CAE technology had emerged at just the right moment, just when the whole product design and development process looked as though it might become unmanageable. CAD/CAE had the potential for coordinating collaboration of larger design teams on projects of immense scope and complexity—as the aerospace industry had already learned.

However, there was an important sphere of influence in the automotive industry that did not exist in the aerospace industry. A rocket ship for traveling to the moon had to be engineered very well. It didn't have to be beautiful, however, to succeed in its mission, as even the least sophisticated automobile did. The minds of people somehow managed to coordinate the disparate worlds of the engineering cubicle and the styling studio—albeit, with difficulty sometimes—but no one had yet addressed the eventual need for computers with aesthetic judgment or, to put it simply, "good taste." Clearly, it would be impossible to leave the crucial aesthetic objectives of industrial designers—the only professionals officially charged with responsibility for a product's beauty—*off-line and outside the CAD/CAE loop.* To include aesthetic parameters, we would have to know how to encode what designers called the "sweetness" of a curve in the ones and zeros of binary computer lingo. How would computers come to understand the welter of descriptive qualities all people instinctively see in the looks of things without knowing why: *taut, clean, precise, fast, active, powerful, heavy?* What's the difference, in ones and zeros, between a line down the side of a car that is taut or slack, slow or fast? We would need to understand these things before we could impart good taste to computers. The most intriguing question: Did they already possess it?

Rational aesthetics was an oxymoron in the lexicon of car designers. No designer that I knew believed that cold, rational computers could conjure the emotionally warm and sweet surfaces that characterized beautiful cars. As recently as 10 years ago, a vice president of design of one of the largest manufacturers told me flatly that "car designers will never use computers." Industrial designers, in general, shared this opinion. Some still do. Nevertheless, that same car company cannot, today, transform its design process from paper to computer fast enough. And computer-aided industrial design (CAID) is rapidly becoming the standard industrial design medium. Graduates with industrial design degrees must have mastered CAID. Graduates who are especially skilled with computers can expect substantial salary premiums.

At any rate, Sutherland's presentation convinced me as early as 1964 that such things would come true—from sheer necessity, if not desire. My bosses at Ford Research thought so, too, and sent me to study the matter at the University of Michigan, where I split my time between the departments of computer science and psychology for the next 6 years.

What I learned there and in the meantime—which I have written about here—pertains to more than car design or product design, generally. The underlying principles I have discovered pertain to that most general "sense of beauty" I have sought to understand. They apply with equal validity to artifacts of any creative field you would consider, whether architecture, fine arts, literature, or even music.

As a plan for reading this book, I suggest that you read the last chapter first and look at the examples of product design pictured at www.delcoates.com. The chapter summarizes much of the book, including my notion that an industrial designer uses just four visual ingredients to cook up the tastiest of designs—and needs no more. Blended ineptly, however, the same four ingredients can produce the blandest or vilest of dishes.

That preview will provide a road map that will make preceding chapters easier to grasp. Chapters 1 through 4 concern the premises of my thesis. Chief among these is the notion that the thousands of products that surround us every day constitute a mass information medium of enormous scope and consequence, both personal and cultural. This being the case, we can appropriately consider product design within the context of information theory. Thus Chapters 5 and 6 put forth some basic principles of information, what constitutes an information or communication medium, and how products qualify as media.

Chapters 7 and 8 propose a theoretical basis for the sense of beauty and why we like or dislike something. Our ability to enjoy beauty apparently depends on our very ability to suffer fear; the senses of beauty and fear stem from the same psychological processes. Indeed, the word *ugly* comes from an ancient Norse word, *uggligr,* that means "fearful." This quintessential expression of "fight or flight" sounds at once like a guttural growl and a gasp of disgust. Chapters 9 and 10 explain the property beautiful products have that ugly ones lack—an attribute called *concinnity.*

ACKNOWLEDGMENTS

I owe my greatest debt of gratitude to Dr. Richard Pew, recently retired chief scientist of BBN of Cambridge, Massachusetts, and formerly professor of psychology at the University of Michigan. As continuing director of the University of Michigan's annual Human Factors Engineering Short Course, he has provided an especially valuable forum for testing and tweaking my thesis by extending the privilege of lecturing there each summer since 1970. The scrutiny of thousands of professional engineers, scientists, designers, and human factors professionals—many with responsibilities for product development—has honed my ideas and methods. Dick had asked me to probe the deeper, less obvious relationships between aesthetics and ergonomics. The connections and comparisons opened crucial insights to me. He also was among the earliest to urge me to write this book. I also want to thank his associate in planning and directing the course, Dr. Paul Green, of the University of Michigan Transportation Research Institute. His inputs over the years also have been valuable.

Others gave me important forums when I served as contributing editor of *ID Magazine* and *Innovation,* the journal of the Industrial Designers Society of America (IDSA). No place equals the classroom for honing ideas, of course. Any teacher owes an imponderable debt to his or her students for the knowledge gleaned from every quizzical frown and every doubting question. I can't list all the students of the past 36 years—my "guinea pigs" at the College of Creative Studies and San Jose State University—who have contributed directly or indirectly to this book.

My students have asked for this book so that they might, at long last, have something more comprehensive and organized than my scattered notes, slides, and blackboard scribbles. Others, representing the business community, have urged and encouraged me to write this book to meet the needs of those who manage the product development process. They include Kazumi Yotsumoto, founding president of Nissan Design America (NDA), with whom I had the pleasure of helping to plan NDA as I advised him on design strategy; Noland Vogt, founder of the pioneering Silicon Valley design consultancy GVO, who hired me as his first director of research; Earl Powell, president of the Design Management Institute; and Lee Jerrell, associate dean of San Jose State University's School of Business. Nissan and GVO provided early testing grounds for my methods and thereby contributed greatly to the content of this book. Lee has been an enthusiastic advocate of good product design for so long that I owe much of my clippings file to remnants of his. He convinced me to develop a design course for MBA students, a task that forced me to address the manager's view of design more fully.

A book about design would be pointless without illustrations. Thus I owe many thank to the manufacturers, designers, and other sources of images. I owe more to some, such as John Houlihan, former director of design at Timex, and Kaoru Hudachek, marketing manager of Design Within Reach, and at IDSA, Director Kristina Goodrich and Cordelia Chu-Mason. My students have inundated me with more offers of help than I could take advantage of, for which I am most grateful. I must mention a few who went especially out of their ways to help me gather and prepare illustrations: T. C. Chang, Chris Frank, Agota Jonas, Maz Kattuah, John Ko, Chris Schmidt, Mike Smith and Misha Young. Where no attribution appears in the credits, you can assume that I created the illustration.

My agent, Sherryl Fullerton, deserves much credit for making this book happen. Thank you, Sherryl, for your enthusiasm and expertise in helping to develop a convincing proposal—and for your thoughtful follow-through thereafter. I thank my book midwife, Carolyn Pincus, for the day-to-day guidance that helped shape it throughout its gestation and for the pushing that led to a timely delivery.

I thank my editor, Mary Glenn, for her faith in the project. I thank Sandra Foster, Jon Tepper, and Bonnie Henkels-Luntz for their assistance. I thank San

Jose State for the gift of uninterrupted time, in the form of a year's leave. And I thank Tomasz Migurski, coordinator of the Industrial Design Program, for assuming the additional burdens my leave entailed; he also provided a thoughtful and useful critique of the manuscript. Another San Jose State colleague, art historian Dr. William Gaugler, provided valuable historic insights.

I owe special thanks to my two friends and colleagues of longest standing, Ralf Schubert and Kirk Smallman, for their reviews and recommendations. Kirk and his wife, Shirley, both editors and professors of communication, helped me solve serious problems of organization.

Finally, I want to express special gratitude to my wife, Betsy, for her valued suggestions and for her patience while I took so much time away from our normal lives.

WATCHES TELL MORE THAN TIME

The Product Sea

1

WE SWIM DAILY IN A SEA of mass-produced products that flood our awareness and shape our minds more profoundly than all the newspapers, magazines, TV shows, and movies we also experience. On the way to a movie you ride in a car that entertains you as surely as the movie will, by reinforcing your values, beliefs, and aspirations—thanks to the expertise of industrial designers, those people who are part engineer but mostly artist, who chiefly determine the way products look. You check your watch, also shaped by an industrial designer, to be sure that you won't miss the beginning of the movie. Your companion admires its beauty, and you feel the warm satisfaction of affirmation. The woman in the ticket booth, adorned with jewelry and eyeglasses designed by other industrial designers, dispenses tickets from a machine shaped by yet other industrial designers. The kid behind the refreshment counter dispenses a Coke from a machine shaped by an industrial designer into a cup shaped by another. Finally, you sit in a seat shaped by an industrial designer and watch a movie filled with images of other products—all shaped by industrial designers. I've only skimmed the surface of this visual sea, of course. I didn't mention the Palm Pilot that might nestle in your pocket or purse. Nor the cell phone you carry with you or in your car. An exhaustive inventory of the artifacts shaped by industrial designers that you witness every day—from the moment your clock radio turns on in the morning until you set it at bedtime—would fill this book and then some.

We don't just passively view the sea of products. We actively seek information from the watches on our wrists, the personal digital assistants (PDAs) in our pockets, and the computers on our desks. In fact, *all* products talk to us, and not just those we think of as information or communication devices—telephones, fax machines, TVs, radios, and watches—but *all* products. They always have. And they always will. Some speak out literally, like those infamous talking cars of a few years ago that announced "Fuel level is low!" and, when you forgot to turn them off, "The lights are on!" High-tech clocks announce the time every hour (or more discreetly, only when asked with the press of a button). Modern high-tech clocks with digital voices represent only the most recent of a long line of vocal timepieces. Grandfather's clocks and ship's clocks have chimed the time for centuries. Few products can match the attention-grabbing ability of an old-fashioned alarm clock's wake-up call.

Although increasing numbers of products come with audible chimes, bells, buzzers, and quasi-human voices, verbal products nevertheless remain relatively rare. Those which do communicate audibly invariably communicate visually as well, and this is what we mean when we say they "talk." In addition to the sound that comes from a radio, labels on its dial and controls also send visual messages. An ordinary watch doesn't tell us the time so much as it shows it, with what amounts to "body language," by silently pointing its "hands" at numbers on its "face." Likewise, the thermometer outside my kitchen window unobtrusively points out temperatures on its clocklike face. The linear thermometer in my medicine cabinet indicates temperature with a slender finger of mercury. Other products communicate in a totally passive and silent manner. A ruler can convey enormous amounts of information. So can a book, which after all, is also a manufactured product.

WATCHES TELL MORE THAN TIME

Your watch always tells you—and anyone else who sees it—considerably more than the time. It is impossible, in fact, to design a watch that tells *only* time. Even the most ordinary watch speaks volumes not only through the customary numbers, hands, and graphic symbols on its face but also by means of every visual aspect of its form—including its shape, color, and texture. In the same way that people embellish their words with a universally understood "body language" of gestures and postures, products also say much more than the words and numbers of their labels suggest.

The appearance of each watch says different things—about itself, its designer, its manufacturer, its era, and the person who wears it. Knowing nothing more, the design of a watch alone—or of any other product—can suggest assump-

TIMEX
10:09.36
1NEW PAGE
TIMEX
TIMEX
INDIGLO
START-STOP
OCT 14
10:09 PM 36
MODE-PULSE
100 LAP MEMORY
IRONMAN
TRIATHLON
INDIGLO
RECALL
'99.10-14
10:09.36
FRI T-1
TIMEX
FLIX
10:09.36

tions about the age, gender, and outlook of the person who wears it. It also conveys implications about its quality, performance, and worth. It suggests as well what the manufacturer deems important.

You could classify a watch according to any other attributes you choose, regardless of how irrelevant some might seem in light of a watch's essential time-telling purpose. The truly remarkable thing about a product's body language (like a person's) is that *it answers any question implicitly put to it by the viewer.* Ask the watches in the figure, "Which of you is heaviest?" or "hottest?" or "coolest?" One will step forward each time to say, "I am." Then ask them whose owner is the oldest, or youngest, or most athletic. Again, one will volunteer each time, "Mine is."

CULTURAL SIGNIFICANCE

The questions a person implicitly asks of a product always pertain to personally relevant concerns and the preoccupations that tend to dominate her thoughts and feelings. (I will use "his" and "her" interchangeably and equally throughout this book.) And every product is a blabbermouth; it has a tendency to answer every question—and then some. The sea of products consequently issues a cacophony, like the sound of an orchestra warming up prior to a concert. However, just as we can detect strains of the first piece on the program, we can discern harmonious strains beneath the visual din of products—expressions of the same values, beliefs, concerns, and preoccupations we all resonate with called *culture.* The form of each human artifact, including that of each product, has moral and cultural significance that reflects not only its creators but also its audience. More important, it reinforces or reshapes the values, beliefs, concerns, and preoccupations of its audience. A time-based culture was impossible before the invention of the clock. We could not have an 8-hour workday without clocks. We could not be late for meetings until watches became commonplace.

Product design is a social enterprise involving product designers and everyone else involved in a manufacturer's product planning and development process. In turn, a product's design reflects and affects its surrounding culture. One icon of our present era, the Braun Aromaster coffeemaker, introduced a seminal design that not only changed the way coffeemakers look but also changed the way people make coffee. Increasingly, office chairs resemble Herman Miller's Aeron chair. Chrysler's minivan changed the definition of the family car in America. The Palm Pilot hasn't merely replaced notebooks and printed calendars; it has changed the ways people plan their time and conduct their lives. Apple's iMac has become such a compelling icon that computers in movies, sitcoms, and comic strips usually are iMacs.

The most popular products, especially those we regard as beautiful, just happen to have the right answers to questions asked by most viewers. As times change and our preoccupations change, we pose different questions. Or having asked the same questions, we selectively perceive different answers. Regardless of what varies—color, shape, size, etc.—the appeal of a product's appearance always depends on the appropriateness of the answers it gives under the circumstances. Thus anthropologists and archaeologists can learn much about what mattered to the people of any culture or era by interpreting their most popular artifacts, the mass-produced products they chose to own.

During the 1950s and 1960s, when gas was cheap, a car shopper would have posed such questions as, "Are you powerful?" Automobiles like the 1960 Chrysler Imperial that symbolically expressed great power—those which were the longest, widest, most massive, and had the biggest fins—appealed most.

Shortly after the energy crisis of the mid-1970s, when gasoline prices shot to unprecedented levels, a consumer was more likely to ask, "Are you fuel efficient?" She would have found that smaller, lighter-looking cars from Japan and Europe also looked better than the behemoths from Detroit. Today, concerns about gas prices and thirsty minivans and sport utility vehicles (SUVs) apparently pale in comparison with concerns about safety and security—and thus the symbolic advantages of large, rugged-looking vehicles that visually promise protection from highway crashes and carjackers.

Being products of the Information Age, designers of high-tech watches and SUVs are under some pressure to make them resemble information devices. Some watches and instrument panels now combine traditional functions with

Braun Aromaster coffeemaker

Herman Miller Aeron chair

1983 Chrysler Minivan

Palm V handheld computer

Apple iMac computer

those of phones, computers, and satellite-guided navigation devices, as well as sophisticated audio entertainment; some require as much attention and intellect to use as a computer or PDA. Today's products are also expected to present active, robust, athletic personas at a time when we are preoccupied with performance, physical fitness, and security. An SUV also must seem safe as a vault in case of accident and as intimidating to would-be villains as an armored tank.

1960 Chrysler Imperial

Archaeologists and historians understand people of other cultures and previous eras by interpreting their artifacts. Cultural elites denigrate the everyday products of pop culture because they speak to large, "unsophisticated" audiences. The rarity of preindustrial, handcrafted artifacts automatically made them more precious and valuable than today's plentiful mass-produced products. However, this is chiefly a matter of market value due to scarcity. Fundamentally, a one-of-a-kind, Bronze Age hair clip is intrinsically no more artistically valuable than a bobby pin just because the machines for spewing millions of them at a time were not yet available and someone had to tediously fashion it by hand. Some day, when only a handful of Apple iMac computers remains, they, too, will become priceless objects of art.

Even before that day arrives, future archaeologists will learn more about us from the iMacs, Timexes, and Buicks they unearth than from what hangs on the walls of art museums. As Philip Nobel so aptly puts it, "Design is the art that is hidden in plain sight." He includes architecture, fashion, and graphic design, which we long ago acknowledged as art forms. The issue here is not so much the value a culture places on its artifacts but rather how valuable they ultimately become as *indicators* of that culture—and the sheer force of their effects on the culture in return. Just as handcrafted tools, weapons, and other useful objects were the most important artifacts of preindustrial eras whose inhabitants depended on them as much as they admired them, people of future eons who wonder about us will appreciate mass-produced products as the *most significant artifacts of the Industrial Age.* They will acknowledge industrial design as the most significant and influential art form of the industrial twentieth century—and beyond. Already, museums all over the world mount comprehensive exhibitions of industrial design with increasing frequency. New York's Museum of Modern Art has done so for decades, and its permanent collection of the work of industrial designers grows almost daily. London and other cities have museums devoted only to industrial design.

By sheer numbers and pervasive dissemination, the artifacts of the industrial age—replicated by the thousands, millions, and hundreds of millions—are bound to influence contemporary cultures more profoundly than the scarce artifacts of previous ages could affect their populations. The meaningful decorations on a Greek urn that communicated and reinforced cultural values could affect only a small group. In contrast, an iMac's equally significant transparent colors span the globe and reach into every region supplied with magazines, newspapers, or billboards carrying Apple's ads. You don't have to own one to be affected by it.

PRODUCTS AS MASS MEDIA

Mass-produced products do not merely overwhelm the fine arts and other less prolific sources of meaningful artifacts. Collectively, they constitute nothing less than a mass medium of such enormous scope and influence that it rivals TV and the movies in its ability to influence mindsets and behavior.

The 1998 rollout of Apple Computer's friendly, translucent iMac had all the hoopla of a Hollywood-produced mass-media event. The iMac stepped onto a stage where a typically American plot played itself out: Steve Jobs, brash kid, cofounds company that creates an industry. Humbled and outfoxed by the big kids, kid loses company. Older and wiser a decade later, comeback kid gets ailing company back and vindicates himself by creating a dramatic turnaround product that heroically pulls the company back from the brink of disaster. He does it largely with superior design.

As evidence of what a product with an innovative, well-crafted look can do, leadership—in spirit, if not in total sales—returned to the iconoclastic company with the unsanctimonious rainbow-colored apple logo that once owned the personal computer business. Amid a "Think Different" ad campaign that berated the greige-box format that had dragged Apple's market share downward during Jobs' hiatus, the cute, plucky little iMac said it all. It was the little computer that could: "I think I can, I think I can—*I can!*"

The Apple faithful (dubbed "Macolytes" by *San Jose Mercury News* columnist David Plotnikoff), who had stuck with Apple through the grim years, at last had something to cheer about. And they had renewed reason to jeer Bill Gates and his northern juggernaut. The buoyant mood recalled the now-classic 1984 anti-big brother Super Bowl TV commercial that launched the original Macintosh by thumbing its nose at the juggernaut of that day, IBM: "Hi!," the cute, friendly little computer said, "I'm the computer for the rest of us."

To the uninitiated masses still bothered by computer phobia, iMac's friendly demeanor beckoned gently, "Give me a try. I won't hurt you!" It went the ultimate measure to allay their fears of guile even further by baring the warm, incandescent glow of its soul through a transparent skin: "See. I have nothing to hide." The iMac quickly racked up 800,000 sales within just 4 months of its launch and rose to become the top-selling personal computer—and arrested Apple's slide toward oblivion. This despite the fact that, like all previous Macs, the iMac cost more than comparable "Wintel" computers (those running Microsoft's Windows operating system on Intel microprocessors). Some 32 percent of iMacs were bought by first-time computer buyers, which history shows are unlikely to switch to other platforms in the future. More important yet, 13 percent were "conquest" sales to people who *switched* to Apple from "Wintels." Such results provide evidence that design contributes to a product's value as surely as performance, quality, or reliability. In fact, superior design is an especially cost-effective means for increasing value and, consequently, profit. Whereas performance, quality, and reliability improvements usually increase costs, superior design needn't cost more than inferior design. Indeed, bad design compounds costs by mitigating potential consumer appeal. Bad design, or even mediocre design, is among the worst and most needless investments.

THE MALL AS THEATER

Michael Wolf says, in *The Entertainment Economy,* "If the '80s and '90s were about 'I want my stuff,' the first decades of the 21st century will be about 'I want to feel better, sexier, more informed, better fed, and less stressed.' " He observes that in the United States, entertainment is already a $480 billion industry (not including TV sets and VCRs) and is growing fast. He notes that Las Vegas, whose only industry is entertainment, enjoyed employment growth of 8.5 percent that "beat out China, Brazil, France, and the rest of the United States."

The "fun" or "frivolity" factor in the design of many recent products, such as DaimlerChrysler's PT Cruiser, lend credence to Wolf's contention that we have become "fun-focused" consumers. We seek entertainment in our products, just as we do from TV programming, movies, Disneyland, and even the news.

2001 Chrysler PT Cruiser

Even without the theatrics that accompanied the iMac's introduction, the product sea is a major component of this entertainment economy. Products provide visual entertainment to the extent that—like all forms of entertainment—they *stir emotions and perpetuate existing values and beliefs by reinforcing them.* Every venue in which products are visible—whether on the streets, in our living rooms, or at our places of work and recreation—serves as a stage.

As both a temple of capitalism and theater, a shopping mall has no peer. It serves simultaneously as a source for material goods, entertainment—and acculturation. I purposely say *temple* because I don't believe that the cruciform plan that so many malls exhibit is entirely coincidental. The plazalike intersection of the axes serves as communal meeting place, just as urban plazas and town squares once did. These spaces often embrace stages that feature entertainment throughout the year, from musical groups to jazzercise competitions. Santa Claus holds court there at Christmas time, as does another quasi-religious figure, the Easter Bunny, during his season.

The mall has become the social gathering place that Main Street used to be, especially for kids too young to drive. It provides almost everything towns used to provide. The latest mall designs promote this impression by trading any sense of unity for the chaos of unplanned village streets. Unlike the city or suburb outside, its tightly knitted, enclosed environment makes it easy to get from one shop or theater to another on foot. In an age of street anxiety, it is a safe haven where violent crime is virtually unheard of. Only when shoppers leave its sanctity for the expansive wastelands of the parking lot do they risk muggers, carjackers, and rapists. About the worst calamity that can befall a mall is news that someone has been attacked in the parking lot.

The pervasive presence of mass-produced products in our public and private lives—abetted by images of them broadcast in advertisements in magazines, newspapers, TV programs, and movies—means that products as mass media have a far larger audience than any other medium of information or entertainment. While we can choose whether to read a magazine or book, enter a museum or movie theater, or watch TV, we cannot entirely avoid exposure to mass-produced products. Even as we watch TV, we watch a TV set. You might flee to the proverbial desert isle, but you'd better not scan the horizon or the sky if you want to prolong the escape from this exposure, lest you glimpse a cruise ship or a 747.

COLLABORATIVE DESIGN

Designers of furniture and relatively simple products, working alone, sometimes enjoy the same exclusivity and personal control over outcomes as sole fine artists do. However, most industrial designers collaborate with ensembles of other designers and people from a range of other disciplines and walks of life, including managers from several levels who exercise control over the design and development process. Like movies, TV programs, and architecture, the design of the typical mass-produced product reflects not just the mindset of an individual designer but that of a microcosm of the broader culture. From the outset, then, a mass-produced product likely reflects the culture it springs from better than any work of fine art created by an individual. The ultimate look of a product might begin with an industrial designer's sketches, but it usually reflects the give and take of several—often many—de facto designers from disparate backgrounds and with different concerns, agendas, and talents. Every individual involved in the product planning and development process, in fact, affects the product's final form—and what it communicates.

This de facto designer is part engineer, market researcher, ergonomist, and increasingly, assembly-line worker and agent who provides sales and service. Consumers and users get involved, too, through focus groups and other means for capturing opinions and suggestions from the marketplace and potential users. The outcome cannot help but be a more inclusive expression of cultural values. While this "design by committee" process distresses industrial designers—the only experts officially mandated to ensure a product's visual appeal—it should delight anthropologists. They are interested not so much in the tastes of individuals as in the norms of popular tastes.

The collaborative design process is never democratic. The highest rungs on the corporate ladder come with the most license and clout. And no one takes greater interest in outcomes or has more clout than a company's founder, chief executive officer (CEO), or board chairman. A product development team would ignore inputs from this individual only at its peril. Regardless of how much well-founded effort a team might have invested in a project, a whim of the CEO can change a product's ultimate design decisively. Both Fords, Henrys I and II, were famous—infamous among industrial designers—for exercising their aesthetic prerogatives. For better or worse, nothing came out of the Ford Motor Company for as long as either lived without enduring their scrutiny and applications of their personal tastes. An increasing number of corporate captains, such as Apple Computer's cofounder and CEO, Steve Jobs, have become vocal champions of good design and get actively involved in every stage of the design process.

Good design or just *design*—often with that capital *D*—has become a hot-button topic in boardrooms and the popular press. Two organizations, the Design Management Institute and the Corporate Design Foundation, now promote the importance of good design and guide manufacturers who seek it. *Strategic design* and *design strategy* are right up there with *branding* among the "buzziest" of corporate buzzwords and are considered crucial to a company's future prosperity. At the same time, consumers now expect and demand good design more than at any time since industrial design's golden age of the 1930s. Articles about industrial designers and their works appear almost weekly in magazines and newspapers. *ID Magazine,* once read only by industrial designers, now appears on newsstands and is bought by consumers as well as design professionals. *Business Week* and *The Wall Street Journal* now cover industrial design regularly. *Business Week* assumed sponsorship of the annual Industrial Design Excellence Awards (IDEAs) given by the Industrial Designers Society of America (IDSA) and devotes an extensive spread to them each summer.

Yet few books about management or marketing mention the word *design,* let alone *aesthetics.* In *The Guru Guide,* by Joseph and Jimmie Boyett, which summarizes the ideas of more than 75 of the most highly regarded management authorities—from Karl Albrecht to Patricia Zingheim, with Warren Bennis, Henry Mintzberg, and Peter Drucker in between—neither *design* nor *aesthetics* appears in the index. Even Max DePree's *Attributes of Leadership* contains no mention of design, even though he was once CEO of Herman Miller, the American furniture manufacturer that has championed good design as the guiding principle of its corporate strategy for over 60 years.

Tom Peters stands out as the exception. He has stressed the necessity of good product design in several of his books and newspaper columns over more than a decade. "Design is it!" he enthuses in *The Circle of Innovation,* in which he devotes at least 22 pages to extolling good design's virtues and stressing its necessity. He begins with this telling prediction from Harvard Business School professor, Robert Hayes: "Fifteen years ago, companies competed on price. Today it's quality. Tomorrow it's design."

THE MEANINGS OF *DESIGN*

Hayes' assertion implies that, until recently, design has had little importance as a competitive edge. In fact, design always has been "it," regardless of whether performance, price, quality, or some other competitive factor is what consumers are preoccupied with at the moment. And it always will. Design subsumes all the other factors by determining the character and worth of each and every one of a product's attributes. Design is not just about looks. If a product outperforms the competition by fulfilling its purposes best or demonstrates better quality, it does so by virtue of superior design. If it delivers more value than the rest by costing less in the bargain, this also follows from superior design.

Design is ambiguous, what Uwe Poerksen calls a "plastic word." Like many old and familiar words that we take for granted, it has attached itself to different concepts, found different uses, and come to mean different things to different people over the centuries. It means one thing to an artist composing the elements of a painting. It means something else to an engineer weighing tradeoffs of function and economy in deciding how many screws to specify for connecting two components. It means quite different things to an architect with designs for a house and a politician with designs on the White House. To the consumer who happens to have none of these vocations or avocations, *design* most likely refers simply to "good looks." This usually is what business leaders mean when they enthuse about design as the newest next thing in competitive edges. *Good design* and just *design*—especially when spelled with a big *D*—are simply codes for "good looking" or, more technically, "superior aesthetics."

DESIGNING AS PLANNING

The French have made the various notions of design easier to put into perspective by retaining two slightly different words, *dessein* and *dessin,* for two quite different concepts, for which English-speakers have only the one. In the broadest sense, the French distinguish between a private side of the design process, visible only to the mind's eye of the designer, and public expressions of it, visible to anyone.

Dessein pertains to the covert aspect of design. Its synonyms include *aim, contemplate, aspire, envision, plan, project, propose, resolve, scheme,* and *speculate.* It is strictly imaginary and intangible. More important, this aspect of design involves a future-oriented frame of mind, often laced with hope and idealism. We are all born designers in this sense, without need of formal schooling. You exercise your intuitive design skills in everything you do, from contemplating what *better* color to paint the living room to choosing the *best* thing to wear. You design next summer's vacation, too, as you consider options of where to go, when to go, and how much to spend—and select from among them. You might have begun a long-range design project by planning your retirement, even if all you have done so far is open a savings account.

Our very survival depends on our design skills. It depends on the ability to continually visualize the future, accurately predict how it will most likely unfold, and act accordingly. Even as you prepare to cross the street, you contemplate possibilities and probabilities and weigh costs and benefits: Do you save time by jaywalking or opt for less risk by walking an extra half block to use the crosswalk protected by a signal? If you opt for jaywalking, you must predict the paths of each vehicle in traffic and plan your crossing to coincide with a suitable opening. Having embarked, you automatically and unconsciously design each step in order to move along the intended path, maintain your balance, and avoid stumbling on the curb.

Expectations about the future arise from a complex mental model of the world, which Kenneth Boulding simply calls the "Image." Built up, elaborated, and refined by all of life's experiences so far, the Image represents all that you know about the world "as it is" and, by extrapolation, "as it will be"—at least for the near term. You predict when traffic will be clear enough to cross the street, based on similar past experiences and the intuitive knowledge of physics you have developed along the way. Except for an occasional surprise, assumptions based on the Image enable you to cope quite well as you interact with the real world from one moment to the next, step by step, day by day, and year to year.

Design, however, is about more than coping; it is also about changing things for the better. Herbert Simon observes in *Sciences of the Artificial* that design involves acts that transform an imperfect world "as it is" into a more ideal world "as it should be." In Simon's view, all of a university's "professional" schools—including medicine, law, engineering, and business, as well as architecture and design—constitute design schools. The faculties and students of these schools should properly study and devise ways to improve the world by applying knowledge mined and crafted by sister schools of science and humanities. More often than not, we imagine that some new or improved product would be the best way to close the gap between real and ideal worlds: a more accurate watch; a safer, more efficient car; a less intrusive surgical tool for repairing knees.

Humans are not alone in their propensity for design. Survival for any organism depends on anticipating the future and reacting appropriately to it. My cat Rex designs when he asks to go out or climbs onto my shoulder for his affection fix and to groom my beard. In both cases, he is striving to improve his world as he sees it. Humans take more factors into account and contemplate more distant futures—the hallmarks of superior intelligence. Design is so characteristically human, in fact, that we might justifiably call ourselves the "design animal." Our superior design ability sets us apart from other species as surely as our superior language and tool-making abilities. Indeed, the most important thing about those other talents, especially tool making, is the extent to which they serve the objectives of our design behavior.

The design instinct entails certain psychological costs. A designer is always disgruntled to some extent because things never seem good enough or as good as they could be. Rex would rather be out when he's in, for example, and in when he's out. And my beard is never clean enough to suit him. Regardless of how good a product is, most users can find shortcomings and imagine how it could be better. And the designers who created it were never quite satisfied with it and wish they could have another crack at "getting it right." When I assign my students an open project, in which they have the opportunity to invent any product they wish, I suggest that they begin by developing "bug lists." True designers have no trouble coming up with endless lists of things from everyday life that "bug" them—each representing a potential new product.

Many of our designs are no more than idle musings that never see the light of day: a dream vacation never taken, a dream house never built. Some, like a politician's designs on a particular office, never become tangible even when realized. A product designer's fantasies and plans, however, are meant to be transformed into very real and tangible objects. This transformation—from ideal to real—defines the design process better than any other characterization.

DESIGN AS VISUALIZATION

The other general notion of design, *dessin,* corresponds to the conspicuous side of the design process. Its synonyms include *sketch, drawing,* and *delineation.* These visible manifestations of the otherwise secret design process enable the designer to communicate to others what is going on inside her head. They also play a crucial role in the ideal-real transformation. They enable what Rudolph Arnheim calls "visual thinking," the chief means by which the designer's mental processes move forward. A sketch of a product concept issues from the designer's imagination. The sketch, in turn, informs—or rather, re-forms and redirects—the designer's imagination. This results in a revised sketch, which, in turn, revises the imagination yet again. This feedback loop, of private and public visualizations, continues throughout the design process. The conspicuous sketch is as important as the covert concept in determining the outcome of the design process. Visualization is so endemic to the design process that schools of design prior to the twentieth century were little more than schools of drawing.

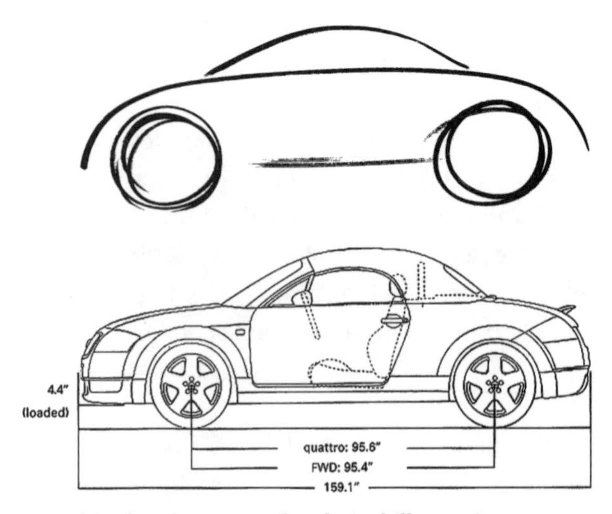

Empathic sketch (top) and technical illustration

Each visualization, from the first conceptual sketch to the final working prototype, amounts to a snapshot of a design caught at some stage of its metamorphosis from imaginary concept to a tangible product. The depictions shown here of the Audi TT Roadster illustrate just two of the broad range of visualization modes used by industrial designers. Neither the empathic sketch nor the technical illustration below it depicts the actual car as fully as a photograph. Neither includes information about color or gloss, for instance. Each one focuses the viewer's attention, instead, on a particular aspect of what the designer has in mind and needs to communicate to others at some point in the process.

A designer normally uses empathic sketches at the outset of the design process to explore and express the aesthetic essence of a design, not its objective reality. The wheels in the empathic sketch are not round, for example, nor even of the same size. And the profile only suggests the actual car's profile. Suitable empathic sketches evoke feelings, emotions, and moods that are appropriate for the product being designed. This particular sketch seems "active," for example, not "passive"—one of the impressions a consumer would expect a sports car to make. Empathic sketching requires unconstrained, intuitive strokes, not thoughtful ones—preferably with a broad felt-tip pen rather than a sharp pencil. A designer might feel emotionally intoxicated in the midst of a fruitful empathic sketching session. Feelings might seem to flow smoothly up from the viscera, down through the arm, and out onto the paper. Thoughts would only get in the way.

Audi TT Coupe and Roadster

Technical illustrations come later in the design process, usually after the concept has been pretty much fleshed out. Whereas empathic sketches are expressive and emotional in nature (typical of an artist), technical illustrations are factual and rational (typical of an engineer). They require a precise drawing instrument and deliberate, careful attention to detail; objective accuracy is all-important. A technical illustration accurately forecasts the product's dimensions, proportions, and other geometric characteristics precisely enough that it could serve as a plan for making the real thing. At the same time, it should not sacrifice the equally crucial aesthetic essence of the empathic sketches that preceded it.

The industrial designer is the only member of the product design team expected to master both empathic and technical extremes. Successfully melding the two in the final product defines the industrial designer's most significant challenge. It is difficult because it requires two quite different frames of mind—emotional and rational—and the ability to switch back and forth between them at will and as necessary.

DESIGN AS EMBELLISHMENT

Dessin, however, implies more than visualization. Other synonyms include *pattern, motif, decoration,* and *ornament.* Before the Industrial Revolution reached its full stride, the closest thing to an industrial design course in one of those old-fashioned design schools mentioned earlier would have involved only superficial matters. Students would practice the design of decorative patterns or motifs for the embellishment of otherwise plain surfaces of fabrics, wallpaper, furniture, and architectural trim, as well as the largely handcrafted products of the period. Superficial decoration of product surfaces did not give way to concern for innovation and the underlying form of products until the advent of the Modern Design movement less than a century ago.

The notion of design as embellishment persists to this day, however, much to the chagrin of modern industrial designers. Art museums still assign furniture and manufactured products to categories with such names as "decorative arts" and "applied arts"—even though most of the items added to their collections since the 1920s have adhered to the norms of Modern Design and its disdain for embellishment. The outmoded view persists on bookshelves, too. If you open the latest book with *design* in its title, likely as not you will find it filled with decorative patterns and motifs for application to anything and everything.

THE URGE TO EMBELLISH

With the perspective of history Modernism's ascetic aesthetic might prove to be a momentary nadir of embellishment. Earlier eras, including those which immediately preceded Modernism (Art Nouveau and Art Deco), were characterized by more embellishment. So is Post-Modernism. Perhaps the pendulum is swinging back toward a more normal median, and what follows next will be marked by even more embellishment.

Modernism's lack of detail might seem unnatural to our nervous systems. There has been speculation that the level of visual complexity we tend to prefer and feel most comfortable with was conditioned genetically by the complex arboreal environment from which our species emerged. If so, we should feel more at home when surrounded by something as visually complex as, say, Art Nouveau. A Modern Design environment might seem unnaturally austere, even when compared with the relatively plain, sparsely vegetated savannahs we later ventured on. After studying photographs of the interior spaces we typically choose to live in, no anthropologist would conclude that we are a culture of Modernists.

The urge of artists, writers, and musicians of any era to embellish their work seems irresistible. What would prose be without adjectives and adverbs to embellish it—to say nothing of poetry? What would music be without an occasional flourish? Even at Germany's Bauhaus School, Modern Design's citadel, designers sometimes spent extra effort to fake a look of effortless, no-nonsense functionalism—just to drive home the point of their Modernist manifesto.

Industrial designers have celebrated the Audi TT as a contemporary exemplar of Modern Design, a design that might have emerged from the Bauhaus if it still existed. Yet various expressions of the bolt-circle motif found throughout

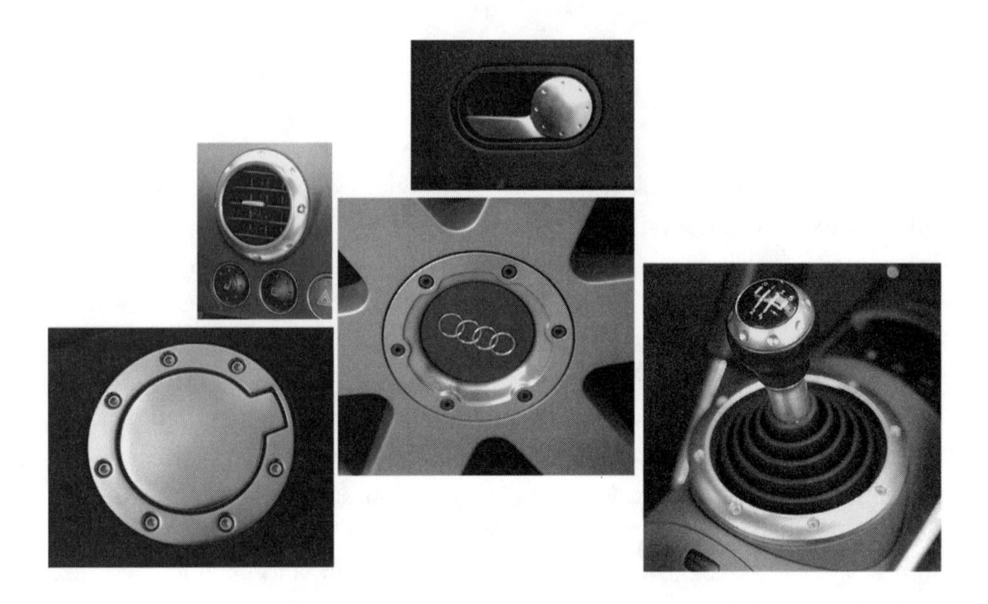

the car qualify as unabashed examples of embellishment. Where the bolts are real, there are more than needed to do the job required of them. Where they are imitations of real bolts or only symbolic dimples, they serve no practical purpose at all. They serve only the aesthetic purpose of *suggesting* that the car's "form follows function" in a strictly no-nonsense fashion—which, of course, the motif simultaneously belies. This tongue-in-cheek irony makes the car truly modern (i.e., up-to-date) with a delightful dose of Post-Modernism.

I hasten to say at this point, if it isn't clear already, that I am not criticizing the TT's design. Nor am I denying the legitimacy of embellishment. I happen to like the TT a lot. I probably would have left off the decorative touches had I designed it, but I am of a generation of designers whose values were thoroughly shaped by Modernism. The team of younger designers responsible for the TT would have been as compelled by Post-Modernism's values to include them.

Critics of embellishment were more convincing during the Modern Design era for practical reasons. Until recently, molding imitation bolts in a hub cap, or even dimples, would have required many more hours of a highly paid master toolmaker's time to machine the molds. This would have driven up the mold's cost and, consequently, the product's price. Today, however, when computers carve molds, they entail no more cost than specifying one color for the plastic instead of another. Thus there is less reason to take the Modernist's hard line against embellishment.

DESIGN AS COMMODITY?

You might assume that a hubcap with decorative bolts embodies more design than one without them. The historic association of design with embellishment reinforces the notion; more embellishment suggests more design. And if you compare the two, the embellished one certainly seems to have more of something. What it has, however, is not more design but more *information* and, as a consequence, more of your attention. I will turn to the subject of information in due course. In the meantime, I want to dissuade you of any notion that design is a commodity that a manufacturer can add to a product to increase its appeal and value, in the way that more gold in a watch's case would increase the monetary worth of the watch.

We cannot adjust the amount of a product's inherent design in the same way we might increase other abstract properties such as performance and durability. In fact, each product comes with the fullest possible complement of design. The decision to leave the bolts off the TT would have entailed just as much design as the decision to include them. If design were a commodity that everyone wanted more of, removing all instances of the bolt-circle motif would diminish the car's design content, along with its appeal. Indeed, the plainer, "less designed" version might cool the passions of some who like it now as it is. At the same time, however, a good many Modernists in the audience—who subscribe to the maxim that "less is more"—would perk up and warm even more to the simpler, "more elegant" design.

It is the *quality* of design that counts. A product cannot have more or less design—only good, bad, or mediocre design. Quality of design depends, in turn, on the quality of the designer's intentions, contemplations, and choices among countervailing options. Thus we come full circle, back to the original notion of design as planning.

It boils down to a matter of attention to detail. As the legendary designer Ludwig Mies van der Rohe noted: "God is in the details." But so is the Devil—in the unattended details. I often remind my students that, regardless of how good their instincts are, they will become better designers as they become better able to answer all my questions during project critiques: What did you *intend* to achieve by giving your product that particular shape? What were you *contemplating* when you put the handle over there instead of here? Why did you *choose* that particular color? I am often surprised at the inability of designers, even pros, to answer such obviously relevant questions. If they can't, they haven't paid enough attention to the details in question and haven't made choices. If the designer does not choose a color, for instance, the product nevertheless will have *some* color. If the designer of record does not take care to chose it, some de facto designer will: perhaps God, with luck; if not, probably the Devil.

AESTHETICS

The new emphasis on design actually corresponds to a shift in emphasis from practicality to aesthetics. The shift has made the industrial designer—the only member of the product design and development team officially responsible for aesthetics—a first-string player.

As originally coined by the ancient Greeks, *aesthetic* simply refers to *sensations* and *feelings—good* or *bad, beautiful* or *ugly.* This original meaning persists in medical and technical terms such as *kinesthetic* (awareness of the body's posture and movement, which arises from nerves buried in joints, tendons, and muscles) and *synesthesia* (the curious phenomenon where perceptions seem to cross sensory boundaries, as when someone literally tastes a color).

Today, however, when someone calls a painting, a song, or a new car *aesthetic*—or, as the British spell it, *esthetic*—they usually mean that it is "beautiful" or "tasteful." This usage presents a problem, however, for it implies that *ugliness*—the presumed opposite of beauty—is *anesthetic* in nature, "devoid of feeling." However, we know that isn't true. The disturbing stir of feelings caused by ugliness often exceeds the exhilarating stir of feelings associated with beauty. Thus we mislead by glossing over *aesthetic* as simply a synonym for *beautiful.*

I will use the original meaning. The beauty of a flower's perfume is aesthetic, but so is the ugliness of a bee's sting. Beauty and ugliness are not opposites, as we normally assume, but almost identical twins. James LeDoux, a neuroscientist who studies fear, contends in *The Emotional Brain* that fear is the prototypical emotion that underlies all others and that all others have evolved from it. Ironically, our ability to feel the pleasure of a beautiful object depends on our ability to be shocked by an ugly one and stunned by fear. Thus expressions such as "stunning beauty" and "drop-dead gorgeous" come close to literal truth. Every beautiful object also contains the seeds of fearsome ugliness.

I often associate perceptions of beauty with the vague sensations of what I can best describe as homesickness. I have always loved to travel and to experience the strangeness of new places and things. By the time I was 11 years old, I had lived on both coasts and five places in between. I experienced homesickness for the first time during that period, when I traveled alone to visit a friend who had moved some 200 miles away. The homing urge I felt after a couple of days approached melancholy, but it wasn't wholly unpleasant. I suppose that my being especially mindful of home reinforced a sense of family and place, as well as the promise that I would return to them soon. I have experienced similar feelings—radiating from my solar plexus—when confronted with stunning beauty. The gut feelings of returning home or when part of a design "finds home" by fitting perfectly into place seem to be of the same cloth. They seem essential to finding pleasure in extraordinary things or experiences.

Little wonder that such feelings should flow from the gut. In *The Second Brain,* Michael Gershon describes an *enteric nervous system* that innervates organs of the viscera—heart, lungs, stomach, intestines, etc. This visceral "brain" rivals the complexity and influence of the more familiar brain in the head. Several thousand nerves connect the "brain" below to the brain above. Thus we are consciously aware of much that happens down there. However, most of the 100 million nerves of the gut's "brain" exist only for the purpose of communicating among visceral organs. They enable and coordinate such essential activities as breathing, circulating blood, and digesting food. They do so automatically and, for the most part, without bothering us with a need to think about what goes on in our gut.

When the gut's "brain" does make itself known, it tends to dominate the head's brain. A gut feeling can overcome the most reasoned thought. Unfortunately, for anyone trying to explain aesthetics, it speaks a foreign language the head's brain finds hard to translate. Although we have all experienced the melancholy of homesickness and the thrill of beauty—from below—they remain difficult to explain, if not inexplicable with the discursive language from above.

MAKING SENSE

A product's originality—its *novelty*—has a great deal to do with its ability to grab our attention, hold it, and start the aesthetic juices flowing. Novelty alone, however, does not suffice to make us like the product. The unique and novel tilt of the Leaning Tower of Pisa accounts for its fascination but not for any inherent beauty. Despite the fact that people flock from all over the world to gawk at and enjoy its unique deformity, it is an aesthetic disaster as a tower. It has a serious deficit of authenticity. It doesn't *make sense.*

We could increase the tower's architectural authenticity and aesthetic appeal—if not its tourist appeal—by straightening it. In fact, we could increase the sense of authenticity and appeal of just about anything by making it, or some part of it, precisely vertical. Even people become more appealing when they adopt upright attitudes, whether with respect to posture or character.

Elegance implies authenticity or fitness, too. In scientific, artistic, and social spheres, it connotes correctness, exactness, precision, neatness, harmony, and simplicity. It is associated with "tasteful opulence in form, decoration, or presentation," as well as with "restraint and grace of style." It is also used simply as a synonym for *beauty.*

The *sensation* of a watch's weight as you hold it is as authentic as it gets. The purely visual impression that it is heavy does not guarantee that it will prove to be heavy when picked up or weighed on a scale. By the same token, a watch that seems accurate, or durable, or expensive to the eye might be none of these things. In each case, the feelings only amount to intuitive hunches—based only on the preliminary visual evidence—short of weighing it, testing its accuracy and durability, and asking its price. As unreliable as feelings often are, they nevertheless enter into every decision, especially those which consumers make. While it might gall our sense of practicality, a semblance of authenticity is often enough. We often buy products on the basis of feelings alone, without ever testing their true weight, accuracy, durability, or any other attribute that might otherwise sway us.

Still we value a truly authentic product more than one that only pretends—when form really does follow function. If someone in the market for a watch expects and wants a watch that is massive, accurate, and durable, he will tend to be drawn to those which *seem* heavy, accurate, and durable—and what's equally important, he will be inclined to ignore those which don't. However, he will especially like those which prove to be true to their promise and turn out to be actually heavy, accurate, and durable—especially where perceived value equals or exceeds actual price. An authentic product seems "worthy of trust, reliance, or belief," just as an authentic person does. It is bona fide, the real thing. However, the concept implies more than "conformity to fact." Authenticity also implies originality. An authentic object is neither a copy nor a counterfeit.

CLASSICS

Some products continue to pleasantly excite people long after the excitement of their newness has worn off. They continue to seem appropriate for their intended purposes. They continue to fit harmoniously into their surroundings, their circumstances, and their times, even though all these "environmental" factors might have changed in the meantime. Put simply, they *make sense* indefinitely.

We tend to call perennial favorites *classics.* But this word does not go deeply enough to explain the special appeal of beauty. We revere classics in large part because they are beautiful. Or perhaps their beauty lies beneath the skin, in which case they are beautiful because they are classics; *ugly classic* is an oxymoron. Some endearing objects, however, even enduring ones, remain humble and never achieve classic status in the fullest sense of that word. In addition, we have diluted *classic* through overuse, misuse, and laxity until it has lost the special air that once surrounded it. Too often, now, *classic* just means "old," "rare," and "expensive."

A true classic is extraordinary, and not just because it has weathered time. It stands above the crowd even though, as time passes, it might come to resemble the crowd more and more—or rather, the crowd comes to resemble it through emulation. A classic's appeal flows largely from its originality. It is not derivative of something else. In fact, we can attribute the very essence of the word *classic* to the strangeness of originality. When something departs from the norm enough to seem original, we have difficulty fitting it neatly into some preexisting mental category or class. That is, we have trouble making immediate sense of it. Since mental classification is crucial to understanding anything, we cannot long endure the mental unease that comes from something that does not "fit," something that we cannot name with finality. We finally overcome the cognitive dissonance by rejecting the misfit; we might call it "ugly" or, more kindly, "ahead of its time." Or we overcome the cognitive dissonance by creating a new mental class—just for it, alone. It thus defines a new class of things, as the Palm Pilot did, and becomes the truest example of a *classic*. Creation of a new class for the Pilot was easy because "handheld computer" was a natural extension of "laptop computer."

The prototype of a watch designed by Nathan George Horwitt in 1947 stands as one of the most eminent icons of Modern Design. It might well be the most

Prototype watch designed by Nathan George Horwitt (1947)

emulated watch of all time. It remains "one of a kind" and the most authentic of its kind by virtue of its originality. It is the standard by which all subsequent "numberless" watch designs have been, and will be, measured.

The single dot at the 12:00 o'clock position symbolizes the sun at noon. Functionally, it reinforces the vertical orientation of the watch's face. The dot is not strictly necessary to the watch's ability to indicate time; our sense of vertical orientation (and where noon lies) is so innate that we could tell time with the hands alone.

By separating the hours of the day into symmetrical halves, the dot expresses the very essence of our time-telling tradition.

Technically, you could still tell time with the Horwitt watch if its dot were not positioned straight up at noon, but it would seem "imperfect" or "out of kilter." Without a precise sense of vertical orientation and symmetry, this watch probably would seem less authentic, less elegant—and less beautiful—than the genuine Horwitt watch. Which one do you like best? The slightly askew arrangement would have adverse practical consequences, too. More time and mental effort would be required to tell the time using it and it would be more likely to provoke mistaken readings.

You can test the relationship between vertical orientation and authenticity quite easily next time you see a picture hanging askew; straightening it will immediately make it seem more natural, more proper, more authentic. If you happen to be involved with product design, the next time a design doesn't seem just right—but you can't quite put your finger on why—try making one or more visual elements seem vertical, especially any that are nearly vertical already but not quite. Chances are good that you will like the design more after the change.

CONCINNITY

None of the preceding characterizations—classic, fitting, original, elegant, authentic—is sufficient, alone, to explain beauty or what we call "good design." Others, equally wanting, have been proposed, including uniformity, order, redundancy, and coherence. I know of only one term general enough to account for any property that helps the viewer "make sense" of a design: *concinnity.* Rarely used for centuries, it nevertheless appears in most dictionaries over 2 inches thick. They typically equate it with "symmetry," "harmony," "neatness," and "elegance." A few cut to the chase by saying it means "beauty" or, more specifically, "studied beauty." From the Latin word *concinnitas,* it originally meant "a skillful and harmonious adaptation or fitting together of parts." Think of an arch whose stones fit so neatly together that they need no mortar to hold their place and prevent collapse. Later, in rhetoric, concinnity was called a "a close harmony of tone as well as logic among the elements of a discourse."

Don't let the double *n* trip your tongue; it's pronounced no differently than if it had only one *n*—kun-SIN-iti, with the accent on *cin.* To the extent that Latin was pronounced like its daughter, Italian, the double *n* might be savored for just a moment longer than a lonely *n.* Spoken naturally, it slides smoothly and effortlessly off the tongue—rather concinnously, as a matter of fact. Its adjectival form, *concinnous* ("kun-SIN-us"), is more sensuous yet; it doesn't merely slide, it slithers.

Technically, the inherent concinnity in the form of an object makes it easier to perceive, understand, and remember. A cube and a sphere, to use just two examples, have lots of concinnity. And you know precisely what I mean by *cube* and *sphere*. Their images come quickly and easily to mind. You probably could draw a pretty accurate picture of either from memory. Even though you are quite familiar with your car or your watch, you would have considerably more difficulty drawing an accurate picture of either. They might have a lot of concinnity for a car or a watch, but no car or watch will have as much as a sphere.

The sense of beauty requires that a thing "make sense." Concinnity ensures that it does make sense. Whenever you see something that is beautiful, you can be certain that it embodies concinnity. By the same token, when you see a product—or anything else—that is wanting aesthetically, its shortcoming always can be traced to a lack of concinnity. Concinnity is in the repetition of rhythmic and rhyming words of a poem that makes it easier to memorize. Concinnity is in the repetitious theme threading its way through a song that makes it easier to sing. The Horwitt watch reveals its concinnity in several ways: The circle has the greatest possible symmetry of any figure; the repetition of circles lends it harmony; and its simplicity lends it elegance, as does the realization that it achieves so much—functionally and aesthetically—with so little. All these concinnous factors make it easy to describe or draw from memory.

I will explain more about the nature of concinnity and how a designer can use it to turn a sow's ear into a silk purse in Chapters 9 and 10.

WHY AESTHETICS MATTERS

Why should people care so much about a product's aesthetic qualities, which, after all, are ethereal and do not necessarily correspond with how well the product works? Why do an increasing number of art museums mount exhibitions of industrial design alongside the works of Michaelangelo, Van Gogh, and Picasso? What warrants all the attention in boardrooms?

The hair's breadth separating beauty and ugliness accounts for all the perplexities that vex marketers, CEOs, and designers who hope to appeal to consumers: the product loved by one but loathed by another, the product liked last year but disliked this year, the product shunned last year but embraced this year, and the holiest of all grails, the product with classic beauty that everyone admires forever.

In the final analysis, aesthetics is about pleasure—or displeasure. And the *pleasure principle* is the fundamental motivator of everything we do: We instinctively seek out things that promise pleasure and avoid those which threaten displeasure or pain. Even when nothing particularly important is at stake, such as hunger or sex, the search for pleasure and avoidance of displeasure are so ingrained that we take our pleasure wherever we find it. Thus a product's beauty often provides reward enough, just as the beauty of a revered sculpture does, or the sound of a brook, or the smell of a flower. The Yankee utility of a Porsche or Rolex watch is virtually secondary to their aesthetic pleasure; the utility of a Ferrari *is* secondary. And they're so much more accessible than works hanging in museums. Even if we can't afford to own any of these precious products, enough of them are out and about that we can at least see and enjoy them more often than the *Mona Lisa*—without the expense of owning them.

What seems a greater willingness to admit our hedonistic leanings lately—even to celebrate them—brings to mind psychologist Abraham Maslow's "hierarchy of needs": Once basic and practical needs such as food, clothing, and shelter have been met, we can turn to more self-fulfilling intellectual, emotional, and aesthetic pursuits. We might imagine a corollary proposition: With assurance that competing products perform equally well enough, last equally long enough, and cost about the same, we can afford to purchase them on aesthetic grounds alone. Furthermore, we are often willing to pay a premium for product beauty. People pay thousands of dollars more for superior appearance in a car, or even a watch, above and beyond either product's practical value. Computers and other high-tech products also fall increasingly into the same category as the cost of their technology diminishes and they become more equal in utilitarian ways.

Eventually, we want to understand why so many products become important to us—especially why we like the looks of one and dislike those of another. We are after nothing less than an answer to the age-old question: "What is beauty?"—and all its perplexing ramifications. Why do you find a certain car (or person, for that matter) beautiful, whereas a friend doesn't? Why does it seem appropriate to wear one watch while jogging but essential to wear another one to the office or out to dinner? Why do tastes vacillate? Why does the new car you loved just last year now bore you? Why does the one you thought ugly last year now seem beautiful?

Design Priorities

2

A PRODUCT HAS at least three audiences—each of which views the product sea through different values and expectations. They include two obvious groups, *users* and *consumers.* Even if, like caterpillars and butterflies, consumers and users usually share the same body, they do so at different times and under different circumstances; once a consumer becomes a user, circumstances and attitudes change dramatically. What satisfied the consumer might not satisfy the user. The largest audience of all—members of the *public* at large who also see the product—consists of millions of men, women, and children of all ages, from all places and backgrounds, who are destined never to be users or consumers.

AESTHETIC VALUE

While practical and economic reasons figure into a consumer's decision to purchase a product—they are the reasons people most often give for buying products—*consumer preferences begin and end on aesthetic grounds.* This means that consumers have more fun, on the whole, than users—who actually depend on a product's practical and ergonomic virtues. Consumers enjoy the bewitching entreaties of charming products trying to seduce them. However, once consumers have metamorphosed into users, their needs, interests, and concerns change forever. And the realities of the new life almost always entail disappointment because the product inevitably fails to meet some user expectations. At home, out of the box, and spread out over the kitchen floor, the "sexy" new computer's components now have to be strung together with a confusing tangle of cables. It has to be loaded with inscrutable software and anxiously coaxed into life. There is no fun in that kind of excitement. Once the long waits on hold for technical support begin, the user could care less about the system's visual virtues. Indeed, if we measure the true beauty of a thing merely by the sum of pleasures we derive from it, as philosopher George Santayana proposed, then the user might begin to wonder why it seemed so attractive in the first place.

Practical, ergonomic, and economic considerations mean nothing to members of the public audience. For them, products they never own or use can have *only aesthetic value.* Products, for them, differ in no significant way from nonutilitarian works of art, which exist for art's sake alone. Aesthetic value affects the thoughts and feelings of those who witness a product—without immediate, practical consequences. For the casual onlooker who sees a watch on someone's wrist or in a shop window, it does not matter whether the watch tells the right time or whether it works at all. It would lose nothing of importance to such appreciators of aesthetic value alone; the inherent aesthetic value of its appearance would remain intact as long as its visage did.

If watches seem to be a trivial example for supporting these contentions, consider modern jetliners. For all their utilitarian importance, they fit into the aesthetics-only category of products for millions of people who have never flown yet who see planes or images of planes virtually every day. Even residents of desert isles are likely to see one flying overhead eventually. More numerous products such as Porsches and Apple iMac computers also occupy this category for most people.

In the end, the aesthetic power of an artifact—and its power to influence—depends on the degree of presence it attains and commands. The iMac might not be more recognizable than Leonardo da Vinci's *Mona Lisa,* which is arguably the world's best-known painting, but it may well rival the second most notable painting, which most of us can't name. The humblest of artifacts, such as the ubiquitous Coke bottle, can achieve iconic status by virtue of sheer numbers or, like the original Volkswagen Beetle, by virtue of endearing qualities and association. Presence can accrue to products of special technical or innovative significance. The supersonic Concorde airliner, of which only 13 have been built during 24 years of service, quickly upstaged the far more numerous Boeing 707 and 747 in aesthetic power. The first, last, and only Concorde crash cost fewer lives than the crash of a single 747 might. However, because of its enormous aesthetic presence, that crash seemed more tragic. While it was indeed tragic for the relatives and acquaintances of the 109 passengers and crew who died, it seemed doubly tragic for millions of people the world over who mourned the demise of the Concorde itself when it summarily lost its legal right to fly.

PRACTICAL AND ERGONOMIC VALUES

Users are compelled to appreciate a product's practical and ergonomic virtues more than any other: As a practical matter, a watch must tell time reliably; as an ergonomic matter, it must be easy to adjust and comfortable to wear. The *practical* implications of product design—under the rubric of engineering—are obvious enough: A product has to fulfill its intended purposes without breaking down, and it has to be made at a cost low enough to yield an affordable price for the consumer, with sufficient profit left over for the manufacturer.

The perspective of the mechanical, electrical, or computer engineer is, for all practical purposes, *product-centered*—even though everyone understands that the product ultimately must serve some user. It has become increasingly important that someone bring an exclusively *user-centered* perspective to the product development team as products become more sophisticated and, in turn, require more of the user (think of the exasperating VCR). The *ergonomist* fulfills this role as the "user's champion."

Ergonomic is another one of those relatively unusual terms, like *aesthetic,* that requires fuller explanation. With the recent popularity of ergonomics—also known by various other names such as *human factors, human engineering,* and *engineering psychology*—many people mistakenly assume that any product with the soft, rounded forms of what one wag called the "blobism" school of design qualifies as ergonomic. However, ergonomics involves much more than just another look or style. From the Greek word for "work," *ergon,* and *nomos,* meaning "laws," it literally pertains to the "laws of work." Thus ergonomists try to minimize the effort involved in physical work, such as lifting or assembly-line work. They also do whatever they can to ease the mental effort of working at a computer or programming a VCR. If you have ever taken a physics course, you realize that work has several physical equivalents—including *energy, heat,* and even those *calories* you try to shed at the gym—all of which come under the purview of ergonomics. In summary, good product design posits three general ergonomic objectives:

- *Minimize nonessential work.* A light laptop computer is ergonomically better than a heavy one because it requires less work to carry it. While the mental effort associated with using it is less obvious than the physical effort of toting it, a confusing display needlessly burns the same kinds of calories with the same result—premature fatigue.
- *Optimize essential work.* Better ergonomics sometimes means dividing work more appropriately rather than minimizing it. A larger, softer handle on a can opener directs more of the user's energy into opening the can and less into gripping the opener.
- *Minimize potentially harmful energy exchanges between the user and the product or environment.* Soft, blunt surfaces on a car's dashboard spread the forces of an accident over a larger area of a passenger's body than hard, sharp surfaces would—just as an airbag does—to minimize the transfer of harmful energy from the crash to the passenger.

THE "ANESTHETIC" IDEAL

The tip-off to superior ergonomics stems not from a product's appearance so much as from its feel when touched or grasped—or more precisely, its relative *inability to produce sensation.* Good ergonomics tends to be *anesthetic* in nature. But remember, *anesthetic* doesn't mean "ugly." You can have a chair that is both beautiful to look at and comfortable to sit on. However, ideally, you would not feel the chair when you sat in it. If you can feel a chair's front edge under your legs—because the cushion has a squared edge or a prominent welt—it could stand ergonomic improvement.

The appearance of a handle is not as important as its tactile and kinesthetic "feel" (e.g., weight, balance, and inertia). Thus the array of mockups of handles considered for Oxo's Good Grips line of kitchen utensils had to be tested by holding them, not merely by looking at them. Ideally, the best handles produce minimal sensation when grasped. They are anesthetic. This means that

OXO Good Grips kitchen tools

they conform to the user's hand as much as possible with gently curving surfaces. This leads naturally to the relatively fat, "blobby" shapes associated with the ergonomic "look."

The handle of a tennis racquet is an exception. Because proper play requires the player to rotate the racquet slightly when changing from a forehand to a backhand grip—without looking at the racquet—the faceted surface of the handle tactilely informs her of how much to shift her grip. The anesthetic standard nevertheless has an important role in tennis. When the ball hits the racquet's "sweet spot," it transmits the least possible force to the player's hand and arm. Thus, when the ball is hit properly, the experience can be virtually anesthetic when compared with a poorly hit ball. Furthermore, the player has an instinctive feel of where the ball will go as soon as it is struck. This "sweet spot" effect is common to many other sports, including golf and baseball.

PRIORITIES

If industrial designers have a patron saint, it is Leonardo da Vinci, the quintessential Renaissance man. Known as the creator of the world's most famous painting, the *Mona Lisa,* he is also known as a prolific inventor and visionary. He blended science, engineering, and ergonomics with aesthetics to conjure the many farsighted flying machines and other concepts for which he is famous.

Following the Renaissance man's example, the ideal product design process rests on a stool supported by three equally stout legs—*practicality* (engineering), *ergonomics* (convenience, safety, comfort), and *aesthetics* (beauty)—that provide steady, balanced support. Shorten one of these crucial legs—or remove it—and you set the product for a fall. As a practical matter, a well-designed watch indicates time accurately enough to meet the user's practical needs. Ergonomically, it is comfortable to wear, and it indicates the time clearly enough to satisfy needs, whether for timing approximate arrival at a meeting or the precise duration of a sprinter's dash. Aesthetically, it must please the owner through his eyes and all other affected senses.

Ideally, the design process would reflect a balanced blend of all three objectives—*as if none had been given preference over any other.* However, the give and take among design team members with different interests and concerns seldom yields true balance. Thus the successful team must agree on a deferential hierarchy among practical, ergonomic, and aesthetic objectives. *In doing so, the team inevitably ends up choosing to favor either consumers or users.* As we have seen, consumers and users have different, sometimes conflicting needs and wants. This fundamental defines what might be called the "marketing dilemma."

Ample evidence from the marketplace argues for an aesthetics-first strategy, especially over the short term. Good ergonomics might yield user satisfaction—with good word of mouth and future sales to follow—but only *after* someone actually buys a product and begins using it. Good aesthetics, however, yields sales and profits right away. Price, quality, comfort, safety, and utility might be foremost among the stated concerns of product planners, designers, engineers—and consumers—as they should be. Nevertheless, no company would knowingly and purposely repel consumers by marketing an ugly product. And no consumer, given a reasonable choice among otherwise comparable products, would knowingly opt for the displeasure of owning one. In the end, no one decides whether to buy a product, or which product to buy, until it looks and feels right—no matter how thoroughly she has consulted friends and consumer magazines.

Corporate decision makers face a dilemma as old as the marketplace: Take short-term profits and stock appreciation—by exploiting fads and fashions of current consumer tastes—or lay the time-consuming groundwork—with sound engineering and ergonomics along with timeless aesthetics—for potentially even greater long-term profits and corporate valuation. There is evidence enough among the experiences of truly prestigious corporations to suggest that the latter course is the best to follow. When the consumer's aesthetic pleasures run smack into the user's needs, it is ultimately more profitable for manufacturers to favor users. Users then broadcast praise to all within earshot. The hapless—and angry—user broadcasts, too, but more loudly. Coupled with discoveries by consumer magazines and other media, faulty engineering and ergonomics can cripple a manufacturer's credibility for decades. Consumers will forgive the occasional misstep of a company with an established reputation for doing things primarily in the user's interest.

Consider a classic example from the automotive marketplace. For many years American automakers installed seats with a cushy "showroom feel" that felt good to consumers in showrooms or on short test drives around the block. However, seats with the showroom feel provided little of the robust support needed for comfort over longer periods. Unfortunately, buyers of these cars didn't learn they were uncomfortable on the daily commute until it was too late. Initially, they dismissed seats of German and Swedish cars that seemed too firm, even hard. Short on showroom feel, they provided instead the firm support, characterizing good "road feel," necessary for hours of comfortable driving.

Even after the superiority of ergonomically designed seats had sunk into consumers' minds, Detroit's design studios continued to shape seats primarily for appearance, with little or no concern for ergonomics. The Advanced Seating Research Department of one company had no responsibility for developing better seats but operated as a profit center. It was staffed by just two engineers with only one mandate: reduce the cost of every seat in every car by just 10 cents every year. Thus, at a production rate of 3 million cars per year, they could increase corporate profits by more than a million dollars. This made the bean counters happy, but not the customers who continued to suffer inferior seating. The industry was reluctant to spend money in other ways that would have improved ergonomics and owner satisfaction. Some 20 years after even the least expensive imported car had reclining seat backs, Detroit's marketing representatives answered queries from journalists about why their products did not by saying, "Our surveys indicate that consumers don't want them." Such misplaced priorities help to explain why the Europeans and Japanese established such a presence in the American market. Such well-established perceptions die hard. To this day, journalists and knowledgeable consumers approach new American cars *expecting* to find fault with their seating while expecting the best from European and Japanese brands.

In the end, road feel—which emphasized ergonomics and user interests—prevailed to the benefit of consumers, users, *and* manufacturers. Long-term ergonomic satisfaction of users yields greater dividends to manufacturers than short-term aesthetic satisfaction of consumers. It follows that an ergonomics-first hierarchy defines the best design strategy for building brand equity and brand loyalty. Specifically,

- Ergonomics must come first.
- Practicality, including utility and functionality, is only slightly less important than ergonomics. What good is a more powerful laptop if it's too heavy to carry?
- *Aesthetics must take the hindmost spot.*

The ergonomics-first hierarchy is the only one that makes sense. Who would want to own and use a beautiful product that did not do the job for which it was intended? And who would want one that performed beautifully but was difficult and uncomfortable to use—or threatened life or limb? Thus the *primary* responsibility of engineers, industrial designers, the CEO, and everyone else on the product design team must be to the user and good ergonomics. This is especially true when, as yet, some companies do not employ ergonomics experts.

Despite this imperative, engineers' job descriptions seldom mention ergonomic responsibilities. Engineering schools normally do not teach ergonomics, and it seldom comes up in regular courses. Many industrial design programs normally include a course in ergonomics, but it often only skims the surface by stressing only *anthropometry* (literally, the "measure of man"). Statistical data on bodily measurements found in anthropometric tables are crucial for determining a product's dimensions, but they alone do not solve all problems of safety, comfort, and usability.

THE VIRTUES OF CONSTRAINT

All but the simplest products have to meet many objectives. A watch's case has to be attractive enough to appeal to consumers. At the same time, it has to be thick enough to contain the works and robust enough to protect them. Its numbers have to be large enough to read under the worst lighting conditions by the oldest possible user. Ironically, each design objective must be expressed in restrictive rather than expansive terms—as rules or *constraints that limit the designer's options*—just as your mother hoped to ensure your safety as a child by curtailing your options: "Don't cross the street!" and "Don't run with scissors!" Constraining your childhood exuberance was more effective at achieving her loving ends than "Play safely!"

The degree of a designer's freedom or discretion depends on how few constraints define and limit the possible outcomes of a design project. A car maker might charge its designers with the objective to create a fuel-efficient car. This objective, however, has no applicable meaning until someone expresses it in more restrictive terms as a criterion or what product designers more often call a *specification* or *parameter.* Thus "fuel efficient" becomes a more tangible and practical specification of "50 miles per gallon." So stated, the objective now amounts to a constraint because it excludes any designs that could not deliver at least 50 miles to a gallon of gasoline.

Generally, the more constraints your mother put on your behavior as a child, the safer you would have been (assuming that you always obeyed her). By the same token, *adding constraints to a design project increases the prospects of a good product.* If 50-miles-per-gallon fuel efficiency were the only constraint, many different designs would be possible. However, without additional constraints pertaining to performance, ergonomics, and aesthetics—to name just a few—many of the possibilities would be underpowered, unsafe, uncomfortable, and probably ugly.

However, the improvement inherent in additional constraints comes at a high cost: Each additional constraint increases the difficulty of the design process exponentially. You can appreciate the scope of the problems by considering just a few examples. When marketing also dictates that the new car should be fast, expressed specifically as a "top speed of 120 miles per hour" and a "0–60 miles per hour acceleration of 7.0 seconds," the number of possible designs shrinks dramatically. Where the essentially contradictory objective of a fuel-efficient but fast car collide, a number of other constraints come cascading down on the designers: If fuel efficiency is achieved with a small engine, then speed must be obtained with a lightweight structure (involving, perhaps, more expensive materials such as aluminum or graphite fiber–reinforced plastics rather than steel) and/or a streamlined shape that requires very little of the engine's power to push the car through the air. Ergonomically inspired federal regulations require crash dummies to "survive" a 35-miles-per-hour crash into a wall. Even if the engineers can sort out and solve these conflicts, they would have other problems in the form of more constraints: Perhaps marketing also wants this new car to be inexpensive (say, $15,000) and, of course, attractive. Such a low-price objective precludes consideration of exotic lightweight materials. Moreover, if consumers are tired of streamlined shapes and prefer the boxy but inefficient looks of SUVs, this conflict presents an almost insurmountable obstacle to the industrial designers as they strive to conjure an attractive look.

It should come as no surprise that product designers sometimes complain that they do not have enough design freedom. This is especially true of an industrial designer who might have come to the work expecting to have the same wide-open creative latitude as his fine arts brethren.

Practical constraints apply even to products such as jewelry, where aesthetic properties are all that matter to the consumer. Jewels come in only certain shapes and sizes, for example. Some precious metals are too precious to meet price targets. The size of a company's furnace limits how much metal can be melted and thus the size of an item. At the other extreme, the physical characteristics of each metal limit how thinly it can be cast without becoming too flexible or fragile.

The best and most experienced designers have learned to actually embrace constraints. They realize, as have other creative people from Michaelangelo to Leonard Bernstein to Orson Welles, that superior art cannot emerge from a constraint-free vacuum. We cannot create anything without tugging at or pushing against some boundary or obstacle that will not give way—short of being shattered. A designer who thinks she can turn her back on one troublesome constraint merely finds herself facing another.

Johan Huizinga makes the case in *Homo Ludens: A Study of the Play Element in Culture* that humans make a game of virtually everything. Unconstrained design, like a game without rules, is no fun—even though you can always "win." The game of unconstrained design has only one rule: When you run up against an unexpected problem (often attributable to an unstated constraint), dodge it by changing the project's objectives or criteria. After changing course, the designer eventually bumps into another problem, of course—and another and another until his trail resembles that of an aimless drunk.

A smart design team avoids wasteful ad hoc adjustments of objective by drawing up a list of all conceivable constraints—thoroughly considered and agreed to *beforehand*—that keeps everyone focused and on track from the very beginning, before pencil ever touches paper. Set down as a list of prioritized objectives and specifications—often called a *product requirements document* (PRD)—the document becomes the final arbiter the team turns to for deciding tradeoffs and settling disputes: "No, we can't use platinum for the watch's case because it has to retail for $39.95."

Regardless of whether constraints are stated explicitly or are simply de facto, *constraints inevitably account for more of a product's ultimate form than the designer's creative intentions.* Consequently, many industrial designers—especially students—complain about a lack of design freedom. So many practical, ergonomic, and legal constraints conspire against aesthetic objectives that these designers do not enjoy product design as much as they had anticipated when they decided to enter the field. This is a naive view, however, born of the notion that industrial designers should enjoy the same freedom as fine artists who work virtually without constraint. Although most industrial designers still graduate from colleges and schools with "Art" in their names, today's industrial designers—as opposed to yesterday's stylists—are expected to embody a more balanced blend of Leonardo's art, engineering, and ergonomics than ever before.

The number of constraints that would thwart the designer grows exponentially with a product's complexity. A designer enjoys more freedom in shaping a drinking glass than a watch because it has only a single part, made of a single material, by a simple process—whereas the watch has many parts and materials subjected to many different processes. The watch is also more likely than the glass to fail; as the number of design constraints increases, so does the likelihood that a product will fail. Charles Eames, the most noted designer of Herman Miller furniture, stressed that the most serious problems occurred where one material had to be joined to another. It represented such a serious problem in the design of his most significant creation, the classic fiberglass shell chair, that only a handful of Herman Miller's employees were permitted to peek behind the curtains to see how its metal legs were attached to the fiberglass shell. Like the Coke formula, it was a carefully guarded trade secret.

There are practical, psychological, and social limits to the number of design constraints and objectives a design team can cope with. These limits, in turn, determine the upper bounds of product quality in every sense—practical, ergonomic, and aesthetic. Thus it is easier to design a good drinking glass, made in a single piece from a single material, than an automobile.

Constrained design isn't always challenging in an enjoyable way, of course. There are limits to the tolerance and capabilities of any designer. Moreover, adding experts to the team to spread the load does not always yield a design to be proud of. Like the infamous committee that set out to design a horse, large teams often end up with camels—and behind schedule to boot.

A product fails because its creators ignored or neglected some design constraint or left it wanting as the result of an improper tradeoff. Even under the best of circumstances, no design team can possibly consider enough constraints to ensure that a product will not fail ultimately in some unanticipated way, short of what should have been a full life. It is hard enough for ergonomists to consider and cope with all that they should—even as they enjoy top priority. However, consider the plight of the industrial designer, trying not to interfere with ergonomic or practical objectives while trying to ensure that the product is also beautiful. Under ideal circumstances, when ergonomics and practicality precede aesthetics, the industrial designer has the toughest job. No wonder many people believe that it is impossible to have a practical, ergonomically sound product that also happens to be beautiful.

Yet it does happen, and with increasing frequency. It will happen even more often as computers come to play an ever more dominant role in the design process.

ULTIMATE CAD

As computer-aided design (CAD) technology matures, the ultimate system will develop the ability to coordinate and guide collaborative design teams of practically any size—while minding enormous numbers of constraints and objectives. This ultimate system will emerge from a fusion of *parametric design technology* and *expert-* or *knowledge-based systems.* With potential for applying and keeping track of virtually any number of constraints and objectives, such a system promises to close the quality gap between simple and complex products and to raise the quality ceiling for all products.

Today's most advanced CAD systems already employ parametric design principles. A *parameter*—which corresponds to some property, attribute, or characteristic of a product (e.g., height, weight, cost, etc.)—amounts to an empty container that can hold any one of several different values or *specifications* that, in turn, inform the product. Each parameter expresses a constraint or objective expressable as a number.

The parametric model of a brick in a computer's memory represents all possible bricks—not just one—because it contains only generic parameters such as "length," "height," and "depth" instead of specific dimensions. The computer cannot draw a picture of it until someone specifies values for all three parameters. Even so, it cannot render a photographic image of it until parameters pertaining to its color and texture are specified as well. Because any number of values can be specified for each parameter, the model represents a very large number of potential brick designs. A more complex parametric model, such as that of a car, which requires many more parameters than a brick to determine its essence, has even greater potential for variety. Put another way, there are many more possible car designs than brick designs.

Not all parameters correspond to numerical dimensions, but they must all be reducible to the ones and zeros used by a computer. You already use a parametric computer system if you write with a word processor. You "fill" the "font" parameter by specifying "Helvetica," "Times Roman," "Palatino," or some other font. You can do this before, during, or after writing—and the selected text changes accordingly. If you do not specify a font yourself, the computer sets it to a default font (if left unspecified, nothing would show up on the screen). You or the computer also must specify a value for the "size" parameter. A "style" parameter lets you or the computer choose "regular," "italic," or "bold."

Any definable attribute or property of a product, regardless of how tangible or abstract, can be expressed as a parameter—as long as it can be expressed as a number and related mathematically to all other parameters with which it happens to interact. In a true parametric model, all parameters must somehow connect. The "volume" parameter for the brick design, for example, would depend on the product of all the parameters already mentioned: length, height, and depth.

Parameters can be fixed (10 pounds) or fuzzy (8–12 pounds). Single-valued *fixed parameters* are precise targets the computer must hold inviolate; multi-valued *fuzzy parameters* are defined by limits within which the computer must stay. A design defined only by fixed variables paralyzes a parametric modeler because it leaves no room for alternative design possibilities. Fuzzy parameters open the way to more potential designs because they free the computer to roam over a range. A fuzzy parameter also can be expressed as an open-ended range ("as light as possible") or as a range bounded at only one end ("no more than 10 pounds").

A third class of parameters, *open parameters,* can be left unspecified, open to any value. A design with even one open parameter has an infinite number of possible permutations, assuming that some other functionally related parameter does not block it. A product with an open "color" parameter could have an infinite number of variants, each of a different color.

Technically, a fuzzy parameter also implies an infinite number of possibilities: A tumbler could have any of an infinite number of volumes in the 300- to 500-milliliter range. Still, fuzzy parameters, with some limits, imply a *smaller* infinity than an open parameter with no limits.

In the ultimate parametric system, the designer will assign parameters pertaining to weight, material characteristics (such as hardness, strength, density, ductility, conductivity), and costs. Any constraint or objective, in fact, that seems important enough to take into account could be expressed as a design parameter. This is important because, as I have already noted, products always fail because of at least one neglected constraint. Thus the designer could take into account even the environmental impact of a particular design for comparison with others. This might mean defining parameters that pertained to the amount of natural gas required to fire a kiln full of bricks of a particular design, as well as the waste and pollutants of the entire manufacturing process.

The expertise of these systems will be realized through the actions of specialized software modules called *agents*—which monitor and manage every practical, ergonomic, and aesthetic parameter behind the scenes with little or no human intervention. Today's computers already employ agents. They scurry about behind the scenes to reduce much of the tedium of routine tasks that otherwise would burden us—like moving, copying, and saving files or searching through databases on the Web. The expertise for constraint-minding CAD systems will be mined from the literature and minds of the best engineers, ergonomists, industrial designers, and other relevant experts.

Specialized agents could oversee anything that mattered: mechanics, ergonomics, aesthetics, costs, and quality control. Virtual product testing already eliminates much of the need for real-life prototypes. Others, such as environmental, social, and economic impacts, are less obvious but no less feasible in the long run—once they can be quantified and functionally related to other parameters. We might expect a new industry to emerge, founded by experts from many fields, that will develop and market specialized agents.

Business schools might need courses in "agent organization." Agents would be programmed to cooperate and negotiate, as we would hope of any team of human experts, to reach the best possible reconciliation of conflicting objectives. Much of what the "weight" agent did to minimize a design's weight, for example, would conflict with the objectives of the "durability" agent, which would analyze the design's structural properties; thinner, lighter parts might not bear up under expected loads. Where conflicts occur, agents would have rules to follow in sorting things out. Planners could establish a pecking order by giving ergonomic agents more clout than aesthetic agents, for example. Being strictly rational beings, unblessed by human emotions, agents would cooperate more reasonably than human committees, always keeping steadfastly to the proscribed objectives and criteria. Ultimate control would remain with the design team, of course. It would have established the rules of the game and priorities beforehand. Agents would decide tradeoffs accordingly but always would defer to the judgments of people when stymied. Like a spreadsheet user, a designer would be free to adjust priorities and parameter values as required and to play "what if" games: "If I give cost a higher priority, will ergonomics suffer inordinately?"

But what a gauntlet! It could be so tortuous that no design could make it through. Of course, some product concepts *should not* get through (the gas-powered leaf blower jumps immediately to mind). In any case, the computer is not likely to ever conjure one design that is optimal in every sense—or optimal in *any* sense. It might well be impossible to devise a product that is at once the lightest, most durable, easiest to use, least polluting, most beautiful, safest, and cheapest. The computer might have to settle for "best possible" in most cases. Heeding Herbert Simon's admonition, we must be content with designing products that are merely *sufficient* in every important respect, not necessarily optimal. All things considered, we will have to settle for products that are light enough, durable enough, easy enough to use, harmless enough to the environment, beautiful enough, and inexpensive enough that we can afford them. However, we should expect that, as CAD systems get better, they will close the gap between what is sufficient and what is optimal.

While ultimate CAD would give practical and ergonomic matters highest priority, a designer might not be aware of the bias because much of the negotiation among agents would take place behind the scenes and out of sight. Ironically, designers would be relatively free to concentrate on matters of lower priority, like aesthetics. For instance, a car designer might design a door simply by drawing its profile on a side-view drawing of the car, with nothing more than its appearance in mind—just as he might on an exploratory sketch. Once he had labeled the line as a "door cut," agents would take over. The ergonomics agent would confirm that it was sized and positioned to meet criteria set for ease of entry and exit. If it did not, the agent would notify the designer and proceed to adjust the line so that it did—in consultation with the aesthetics agent to ensure that it harmonized as well or better than the original with the design's other lines. The "door" agent, which embodies the company's cumulative door engineering expertise, would commence designing a state-of-the-art door. It would consult with agents who knew about hinges, latches, window-lift mechanisms, and other paraphernalia associated with doors. An especially clever agent, the "weatherseal" agent, would begin wending its way, wormlike, through the interstices between the door and the door opening, laying down a design for the best seal the company's experts know how to make. In ensuring that neither rain nor air would leak through the real seal, it would intelligently change the seal's form as it went from the door's bottom, up the back and around the latch, over the window area, and down the front edge past the hinges. Agents would be carrying out thousands of such specialized design tasks simultaneously throughout the car's design. They would be taking care of everything from sizing the engine, transmission, and suspension (already designed parametrically and digitally stored "on the shelf," as it were) to designing headlight and taillight lenses to meet federal lighting standards. Meanwhile, engineers would not sit idly by. They would continue to research, design, develop, and test alternative designs for all these subsystems and elements—in ongoing efforts to improve their technology. As things improved, they would update the expertise of the appropriate agents accordingly.

COMPUTERS WITH GOOD TASTE

I am concerned here with the question of whether products can be beautiful enough, given all the practical and ergonomic concerns that claim priority over aesthetics. I will present a body of quantifiable aesthetic principles and methodologies that make it possible to bring industrial designers and aesthetics on-line and into the full scope of the CAD loop of the future. I have developed and honed these principles and methodologies over more than 30 years of teaching them—mostly to industrial design students but also to students of engineering, ergonomics, and business. I have applied them in the real world, as well, to real products as a practitioner and consultant. They do make tangible the prospects for computer agents with "good taste" that will leverage the design process aesthetically. Even before such computer systems become reality, they make it easier for professional and student designers to achieve good design—faster and more confidently than ever before—with more predictable results in the marketplace. They can make matters of liking and disliking and what philosopher George Santayana called the "sense of beauty" more understandable to anyone who is simply fascinated by such compelling mysteries.

Because I have been concerned primarily with the prospects for using computers to analyze and optimize product aesthetics, I was drawn naturally to see matters in terms of information theory; wherever there is form, there is literally in*form*ation. This perspective made sense, too, psychologically. The nervous system is, after all, nothing more nor less than a seeker, gatherer, processor, and accumulator of information; information is its only resource and commodity. Information also provided the touchstone for comparing ergonomic and aesthetic issues. Thus, while the book is ostensibly about aesthetics, I have much to say about ergonomics in the context of aesthetic issues.

Future "constraint minding" CAD systems will enable designers to explore hundreds of concepts—not to say thousands or millions—with only minimal attention to the welter of constraints entailed by optimal product design methods. With a computer's ability to race ahead and try millions of minute variations and combinations in the time a designer would take to ponder a single concept, it will present the designer with serendipitous surprises—all feasible within the bounds set by the "constraint field" and sorted out by agents. Each surprise will tend to stimulate yet more trains of thought and more creative permutations.

Returning to the example of the drinking tumbler, a parametric modeler would permit the designer to arbitrarily change a design's height (a fuzzy parameter) to check out the appearance of a taller or shorter version. She would be limited to a range of height-to-girth ratios bounded by parameters pertaining to "stability" and "convenience": If the tumbler was too tall, it could be knocked over easily; if it was too short, drinking from it would be like drinking from a saucer. Whatever the case, the designer need not be concerned with volume; the parametric modeler would take care of the details automatically by adjusting enough fuzzy and open parameters—such as the one for girth—to keep the volume within desired limits. And it would reflect the results with a photographic rendering of the tumbler on the screen. The computer could save the original design and any other variants developed along the way so that the designer could make her final selection from a side-by-side comparison of all the candidates.

With the means in hand for analyzing and adjusting the mix of aesthetic factors, computers will have the "good taste" needed to help designers fit products aesthetically into particular market niches and to maximize their appeal to any given group or culture. They will even make some products especially to suit the tastes of particular customers.

Daimons, Zeitgeists, and Icons

3

I FELL IN LOVE for the first time when I was just 4 years old, in the back seat of the family's Hudson Terraplane. Perched on the driveshaft hump so that I could peer over the front seatback, I first saw the object of my affection as she slunk across the next intersection—a sleek, black vision of stealth, power, and elegance. Gone in an eye blink and leaving me with only a dazzling memory. I blurted: "What was that!?" Incredulously, my father hadn't noticed. I pleaded with him to turn and chase her. As we finally pulled alongside, he said, with a hint of awe in his own voice, "Oh, that's a Cord."

The car I fell in love with that day was a 1936 Cord, made by the same company that made the revered Duesenbergs that inspired the accolade, "It's a

1936 Cord

Duesy!" The Cord eventually became an American icon and one of only a handful of automobiles to earn places in the most prestigious collection of all, that of New York's Museum of Modern Art (MOMA). Its designer, Gordon Buehrig, remains to this day arguably the greatest American car designer.

CONSUMING PASSIONS

As in all cases of love at first sight, the scene and its associated emotions etched themselves into my memory so indelibly that I recall them effortlessly and in lucid detail to this day. According to my dictionary, *love* is not too strong a word for characterizing the experience: I had "a feeling of intense desire and attraction," "a deep, tender, ineffable feeling of affection." The attraction expressed itself, quite literally, as a compelling urge to move close enough to caress it not only with my eyes but also with my hands. My dictionary goes on to say that love involves "a sense of underlying oneness" and a desire to possess it, to own it. Coming to *own* something we desire exceeds the legal concept of physical possession. It involves something more akin to emotional integration or assimilation. I wanted to embrace it, to merge with it, to make it not only mine but a part of me.

And I was only 4! I knew little about cars in any practical sense but already had enough "emotional intelligence" (to use Daniel Goleman's term) to know, understand, and appreciate this particular car aesthetically, perhaps as fully and richly as I do today. Apparently, the sense of beauty does not depend so much on the accumulated experience that comes with maturity but rather on something more innate that is present perhaps even at birth. I was already capable of the "consuming passions" characteristic of both lovers and consumers.

I am reminded at this point of the belief held by some, more "primitive" cultures than ours that by consuming an animal—or another person, in the case of cannibals—the consumer assimilates the spirit of the consumed as well as its flesh: food for the body *and* the soul. Thus we call shoppers "consumers" and assess them by their "tastes." And products have "souls."

DAIMONS

The eminent Swiss developmental psychologist Jean Piaget concluded in the 1930s that young children distinguished living from nonliving things simply by their ability to move independently of outside intervention. Accordingly, cars, watches, and other "animated" products do seem more alive than refrigerators and other, more stable products. And they evoke stronger emotional reactions and attachments. More recent thinking has discounted Piaget's hypothesis as too simplistic, but not the notion that children do perceive life forces in some artifacts, especially the ones that seem to move of their own free will.

Sherry Turkle, a psychologist and professor of the sociology of science at the Massachusetts Institute of Technology, finds that 5- to 10-year-old children make relatively fine distinctions and qualify their judgments of just what being "alive" means. She and her research assistant, Jennifer Audley, have studied how children interact with what Turkle calls "relational artifacts" that not only move independently but also behave as though they have intentions, preferences, and other cognitive and affective capacities.

Relational artifacts now available in the marketplace include an electronic doll developed by Hasbro that mimics human expressiveness. Virtual pets, such as Tiger Toys' Furby, and robotic dogs from Tiger, Sony, and other toy makers react to people and to each other. In Japan, where people openly acknowledge the spirits of animals and products alike, often with reverence, the robotic dogs have been runaway best-sellers. Consumers pay as much as $2500 for them, and they aren't buying them just for children. Panasonic reportedly has in development a robotic cat to keep company with elderly people who live alone. When Turkle and Audley ask their subjects whether the toys are like real pets or if they are at least alive, they respond that they are alive—but not in the same way people are. They are alive in a "Furby kind of way."

We don't outgrow the habit of perceiving a Furby kind of life force in toys—or other products—just because we grow older and more sophisticated. Robert Persig wrote eloquently of the spirits that haunt products in *Zen and the Art of Motorcycle Maintenance:*

Each machine has its own, unique personality which probably could be defined as the intuitive sum total of everything you know and feel about it. This personality constantly changes, usually for the worse, but sometimes surprisingly for the better, and it is this personality that is the real object of motorcycle maintenance. The new ones start out as good-looking strangers and, depending on how they are treated, degenerate rapidly into bad-acting grouches or even cripples, or else turn into healthy, good-natured, long-lasting friends.

Ancient Greeks called the spiritual essence we sense in each person or object its *daimon,* from which we derive the unpleasant notions of "demon" and "devil." The daimons of Greek mythology, however, were not necessarily evil or mischievous. Like Greeks and their gods, daimons could be good or evil or anything in between. Generally, the daimon of Greek mythology acted as a "friend," "archfriend," or "genius" that advised and guided a person through what, today, we prefer to call an *inner voice* or *intuition.* (*Genius,* in this sense, does not pertain just to the likes of Albert Einstein. The roots of the word have more to do with "being born" or "coming into being," something we all have in common with idiots as well as rocket scientists. Genius is very close to the Arabic *jinn*—from which we derive *genie*—which probably comes closer to the intended meaning.)

In our own culture, the "printer's devil" was a young apprentice who helped the master printer by taking care of tedious chores surrounding the typesetting process, such as presetting the many *ifs, ands,* and *buts* so that the master would not have to. The printer's devil performed as an *agent,* like those special pieces of software mentioned earlier. In fact, computer programmers call them *daemons* (using the Latin spelling). Finally, there is no better accolade for today's harried office worker than to say that he or she "works like a demon." Sometimes we expect and want something less than angelic. An occasionally revealed mischievous streak can make a friend or lover more fascinating and enjoyable to be with.

No one experiences the daimonic spirits of artifacts more profoundly than the designers and artists who create them. Michaelangelo told of freeing living figures that came from blocks of marble. Novelists tell of their characters taking over the plots of their books. Nothing in my own experience as a designer exceeds the exhilaration of sighting down the side of a full-size clay model of a car and stroking its flank with my palm; the clay is veritably warm with life. I cannot watch an assembly-line robot without seeing a huge insect going through its simple-minded, mechanistic paces. Some kindred spirit draws me to it as surely as the spirit animating my cat Rex.

I sense in the demeanor of Apple's iMac the same friendly and willing agent that animated the original Macintosh of 1984. The slightly upturned face and supplicant posture of each joyfully anticipates an opportunity to do my bidding.

Nothing can exceed the emotional presence of a self-propelled vehicle, which shares the ability to move—apparently of its own will—with living organisms. We can be quite passionate in our feelings for mere objects. When I was 10 years old, my father surprised the family one day with a new car. Although we were excited and pleased to have a new car, my throat grew tight as I fought back tears over the loss of our old car, the one I had known and grown fond of over most of my life. My mother couldn't stop her own tears. "It was," she sobbed, "like losing a member of the family."

Friendliness does not necessarily account for attraction. The friendly assemblage of headlight "eyes" and other "facial features" of Chrysler's Neon are inherently no more attractive than the sinister scowl of the Dodge Stratus' "squinting" headlights. Just as a wild beast grabs more attention than the family pet, the Stratus' menacing, hunkered-down posture provokes more excitement, attention, and fascination than a car with a more friendly, benign demeanor. If the emotional underpinnings of beauty are close in character to those of fear, then the fearsome Stratus might be not only more memorable but also potentially more beautiful than the "friendly" Neon in the long run—for the same reasons that we admire and respect the fearsome tiger more than the friendly family cat.

A person's daimon is no more rooted in reality than a product's. Both are nothing more than ghostly figments of the imagination, due to the same mechanisms of mental illusion. When we describe an acquaintance as

Dodge Stratus R/T Coupe (top) and Dodge Neon

"friendly," we stand on no firmer ground with respect to reality than when we characterize the iMac or the Neon as "friendly." We sense the distinctive daimon in each product's "demeanor" by virtue of the same perceptual instincts and habits that lead us to perceive a person's character in their facial expressions, gestures, and body language. In fact, we perceive daimonic spirits in everything, even the most nondescript stone, because we are instinctively compelled by our natures to interpret all things—natural or artificial—as caricatures of fellow humans. This is especially true if, like cars or watches, they happen to move as if they had wills of their own.

Taking advantage of primary responses to caretakers and other humans, the nervous system—which always prefers the easiest way of sorting things out—probably takes advantage of what psychologist James Gibson dubbed "affordances" in responding to artifacts. Affordances are fortuitous "freebies" of evolution. As an example, our primate ancestors probably evolved opposing thumbs because they were better for grabbing tree branches and enhanced treetop travel. The improved grip, however, also "afforded" them greater dexterity for tool handling. With further evolutionary refinement, the improvement affords us the means for crafting precision watch mechanisms and performing microsurgery—fortuitous, accidental benefits.

The ability to develop emotional, sympathetic bonds with our caretakers—and they with us—early in life has obvious benefits. As helpless infants, we are totally at their mercy and depend on them to like us enough to look after our every need. This ability or tendency seems so basic that it probably has genetic roots that hardwire the nervous system in preparation for bonding. The affectionate exchange between a parent and infant seems to provide psychological and physical benefits to both. The behavior and benefits carry over to other living beings. People apparently live longer, healthier, and happier lives when they have pets to keep them company.

Instead of devising new and special ways for perceiving and responding to artifacts, the ready-made ways of dealing with living things are sufficient and there for the using. Infant monkeys, deprived of their natural mothers, will bond to surrogate dolls that resemble the real thing only slightly. Children derive comfort from dolls and Teddy bears they take to bed with them after keeping company with them all day. The old means might not fit the new application perfectly. Technically, the imperfect fit in such cases "fools" us into seeing life where there is none, but apparently with no risk to survival serious enough to warrant changing them. Indeed, affordances probably enhance survivability, or they wouldn't take hold and persist in the first place. Thus, perceiving a "Furby kind of life" in artifacts might improve chances of survival in some unknown way, or the tendency wouldn't have persisted. Or, if unimportant to survival, this particular affordance might persist simply because such illusions are harmless and afford occasional pleasure. Thus we willingly suspend our disbelief and go along with the illusion.

EMPATHIC EXPRESSION

Daimons make themselves known to us by virtue of *empathic expression,* a technically more general notion than *body language.* It incorporates not only the gestures, postures, and other nuances of body language but also every perceivable aspect of form—including shapes, colors, and textures. *Empathy* (literally "feeling into") looks and sounds a lot like *sympathy* (when two people share the "same feeling"). However, Robert Vischer, the German philosopher who coined the term in 1873, wanted to distinguish the special category of feelings we have for nonliving artifacts (empathy) from those we have for other people (sympathy). While the feelings of both are alike, the ability and tendency to have such feelings for artifacts—as though they were alive—certainly has significance worth special attention and a name of its own.

The automatic urge to smile when you see another person smiling—and to simultaneously *feel* the happiness you presume the other feels—is the "same feeling" experience we call *sympathy.* A photograph of a smiling person produces the same effects, including feelings of happiness.

Volkswagen's New Beetle expresses a "cheerful," "perky" demeanor. Technically, however, we know that the feelings it evokes in the viewer cannot constitute sympathy. We cannot have the "same feelings" as a car or a photograph of one. No product or image has a nervous system, so they cannot possibly have feelings with which to resonate. The only feelings involved must start and end within the viewer. Nevertheless, we all have an instinctive predisposition

Volkswagen New Beetle

to assign a cause or source to every effect and affect—in this case, to the New Beetle. We cannot help ourselves. Compelled by the logical ways of our nervous systems, we conclude that the car is the only possible source of the feelings because they occur precisely whenever it enters the scene.

For several decades after Vischer introduced the concept of empathy, art critics, art theorists, and art historians relied on his definition to explain the emotions, feelings, and other sensations stirred up by works of art. However, as Ellen Dissanayake notes in *Homo Aestheticus,* empathy lost its solely aesthetic connotation when, during the 1960s, psychoanalysts appropriated it as a descriptor of the human condition in general and transformed it into a sort of "supersympathy." According to this interpretation, a person feeling empathy for another person would sense not only their feelings but also the whole gamut of their psyche, including needs, hopes, concerns, and aspirations. It was a richer affect than mere sympathy. Whether we adhere to Vischer's distinction between empathy and sympathy or more recent interpretations does not matter. People form emotional bonds not only with pets and other animals but also with nonliving artifacts.

PRODUCT SEMANTICS

Philosophers have pondered and argued about the meaning of *meaning* for eons. Perhaps they'll never resolve the issue. For practical purposes, however, the meaning of a word expressed linguistically, or of an object—expressed empathically—amounts quite simply to the constellation of thoughts, feelings, urges, and other mental figments of which it literally "re-minds" the viewer. On seeing a body of water—or a picture of water or the word *water*—the meaning of water (its daimon) instantly floods the mind as a confluence of thoughts, feelings, and urges that have accumulated in memory as a result of each and every past encounter with water, images of it, or symbolic representations of it: its "coolness" while drinking it, its "wetness" while bathing or swimming in it, its "smoothness" while looking across a calm pond, the "violence" of a stormy sea, or the "murmur" of a rain shower.

Like most people, you probably think of water as "wet" rather than "dry." You probably also think of it as "cool," not "warm"—unless you happen to be soaking in a steaming tub. Most people also probably would agree that water seems generally "clean" rather than "dirty," "smooth" rather than "rough," and "blue" rather than any other color because most significant bodies of water are reflecting the sky when we see them. This continually reinforced linkage between water and "blue-ness" probably also explains why we think of blue as a "cool" color. If we had gazed out over Martian lakes for our whole lifetimes, under that planet's reddish sky, we might well regard red as a cool color. Or maybe the issue would be the paramount enigma of Martian philosophy because redness also would be inextricably bound with the notion of fire and heat as it is on Earth.

Water also can mean different things to different people. The smooth, clear, and cool semblance of a lake might seem "tranquil" and "rational" to most people. However, it could seem quite "agitating" and "emotional" to someone who nearly drowned as a child. Such a person might grow inexplicably anxious at the sight of any body of water larger than a pitcher. It follows that someone's taste for water—how much he likes it or dislikes it—could be explained by the relative pleasantness of its meaning to him.

Constantly reinforced semantic links between concepts—such as "blue" with "cool" and with "water"—become the dense core of a concept's meaning. A lake does not glow with the reds, oranges, and yellows of sunsets often enough to fracture or reshape the core. Indeed, the very strangeness of water and sky that glow warmly accounts for much of a sunset's charm. The redder and warmer the colors become—the less waterlike and skylike they become—the more interesting, exciting, and charming the scene becomes. Artists take advantage of such contradictions of meaning to energize their works aesthetically. Product designers absolutely depend on them.

THE SEMANTIC DIFFERENTIAL

As director of the Institute of Communications Research at the University of Illinois, Charles Osgood sought a practical, objective method for measuring the differences among meanings of words and concepts from different languages and cultures. The now classic semantic differential survey instrument he developed as a result of his research is deceptively simple and elegant in light of its proven effectiveness. It turned out to be the most effective means I have found for interpreting and comparing the empathic expressions of products and for making portraits of their daimons.

A semantic differential survey is simple to assemble, implement, and analyze. And it paves the way for optimizing a product's aesthetic qualities. A survey typically involves just 15 to 30 pairs of opposing adjectives—such as good-bad, passive-active, powerful-weak, and emotional-rational—placed at opposite ends of bipolar scales. It can be administered in printed form on paper or, better yet, with a laptop or handheld computer programmed to present the scales randomly one by one. Surveys also can be conducted over the Internet or via local-area networks. Subjects participating in a survey evaluate a product's design simply by marking each scale nearer one end or the other, depending on which adjective seems to describe the product's appearance best—or in the middle, if the product seems neutral or neither word seems truly appropriate. Subjects typically spend no more than 5 minutes recording their judgments.

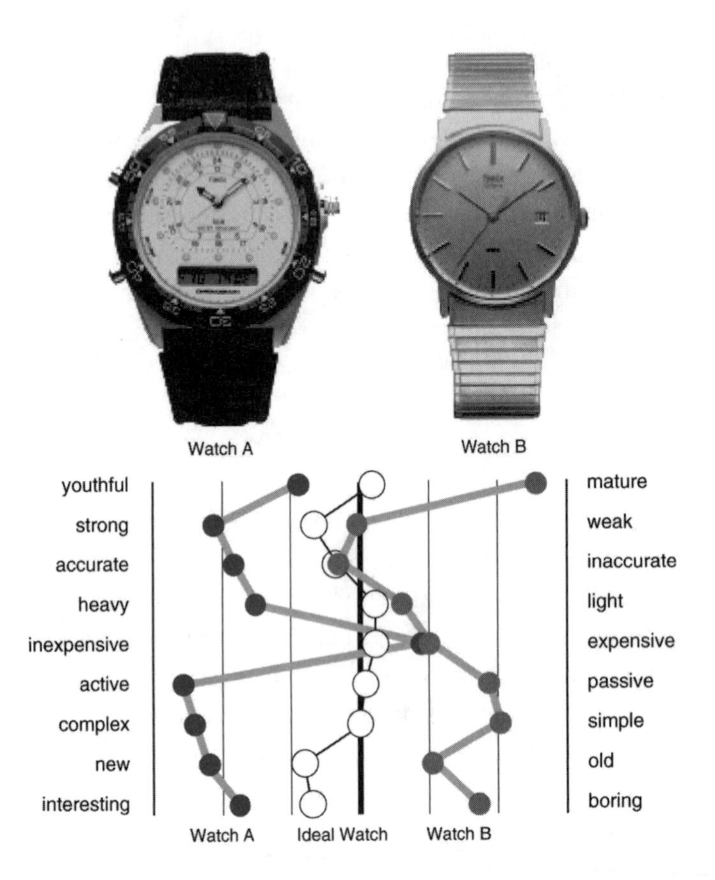

Averaging judgments yields a semantic profile of the product, as shown here for two watches. Watch A and Watch B have decidedly different daimons, as evidenced by the distance between their profiles. Except for their apparent costs, as indicated by the inexpensive-expensive scale, their characters seemed quite different, on average, to the subjects participating in the survey. Watch A seemed stronger, more accurate, and heavier than Watch B, for example. Watch B seemed more mature and simpler than Watch A.

The semantic profile of a product by itself says nothing conclusively about how much consumers might like its design. While the judgments are aesthetic in nature—based on gut feelings—they do not amount to direct expressions of preference. Preferences depend on how closely either watch's design coincides with a particular viewer's values and expectations with respect to watches in general. If accuracy were paramount to someone looking for a new watch, then Watch A might seem most attractive. Watch B didn't seem inaccurate to the group (they judged it closer to the "accurate" end of the scale); it just didn't seem as accurate as Watch A. Watch B likely would appeal more to someone who wanted a light, simple, and relatively mature-looking watch.

The ultimate significance of these semantic profiles comes from knowing how closely they resemble the profile of a hypothetical ideal watch that exists only in the consumer's mind. Because Osgood's methodology works as well for imaginary concepts as for real objects, a third profile was created by also asking the subjects to imagine the characteristics of "the ideal watch" and indicate where it would lie on the various scales. A designer could improve the design of Watch A or B—or any other watch—by altering its design in ways that shifted its profile closer to that of the ideal watch. This could be done by simply adjusting the amounts of information and concinnity inherent in its form. I have already introduced the concept of concinnity. I will describe it more fully in later chapters, as well as what I mean by "information" in this aesthetic context. Meanwhile, I'll just note that decreasing Watch A's information and/or increasing its concinnity would make it seem simpler. This would increase its appeal to this particular group of subjects by moving its profile closer to the ideal. Conversely, increasing Watch B's information content would improve its appeal by making it seem more complex. It also would seem more complex, and technically more ideal, if it lost some concinnity. But it's generally bad policy to decrease concinnity.

Note that, although both watches might be revised enough to bring them into precise alignment with the ideal and with each other, they would not necessarily look alike. They could still look quite different. This alone proves that there is never just one suitable solution to a design problem; the possibility of a beautiful product always exists, even when aesthetics take a back seat to engineering and ergonomics.

Survey participants don't have to understand why one product seems relatively more emotional, for example, or more rational than another; they need only feel it intuitively enough to place their mark. As it turns out, people do this quite readily and, in groups, with considerable consensus. In choosing the scales for a semantic differential survey, two things seem especially remarkable:

- People can relate any adjective pair to any tangible object or imaginary concept—regardless of how relevant the associations might be. Adjectives of greatest relevance yield more useful information, of course; hot-cold scales should appear in surveys of refrigerator and stove designs. Similarly, an accurate-inaccurate scale should appear in a survey of a watch's design, and a car's profile should include a fast-slow scale. In effect, however, a product empathically answers any question put to it, regardless of how germane it happens to be. So a hot-cold scale yields guidance for the design of clocks and cars, too. While pairs such as interesting-boring, expensive-inexpensive, heavy-light, and sweet-sour have little or nothing to do with a watch's practical ramifications, they nevertheless affect its consumer appeal. As we will see later, a rational-emotional scale provides an especially useful glimpse of any product's daimon and a way for categorizing all products aesthetically.
- Results of surveys reveal remarkable degrees of consensus, even when they involve scales that seem relatively irrelevant at first glance. Even though the apparent temperature of a watch might seem irrelevant to how good a watch it actually would be, people will nevertheless agree on just how hot or cold it seems—and how hot or cold it should seem.

ZEITGEISTS

Your *preoccupations*—the myriad things generally on your mind at any given moment—also affect the meanings of things. If you are thirsty, this fact will shape water's meaning at that moment in a particular way. It will likely bend your thoughts, feelings, and urges toward the quenching possibility of a drink rather than an imaginary swim. Similarly, a swimming pool has a different meaning on a hot, sultry day than in the middle of January. A person who once nearly drowned might enjoy sunsets from the shore but find it difficult to enjoy one from the deck of a boat because memories of the near-tragedy come to the fore.

Some preoccupations are more general and more persistent. Someone who suffers from claustrophobia has a persistent concern about confining spaces that might lead him to prefer cars that seem very roomy inside. Preoccupations that are shared by many people at the same time are largely responsible for the ebb and flow of fads, fashions, trends, and styles. When the population of practically the entire industrialized world suddenly became concerned about fuel shortages and prices during the energy crises of the 1970s, that coordinated concern precipitated a sudden trend toward smaller, more aerodynamically efficient cars. As American concerns for power, safety, and security eventually supplanted concerns about gas prices, tastes swung back to horsepower, bulk, and mass—epitomized by the ubiquitous sport utility vehicle (SUV).

When a particular mix of ideals, hopes, fears, and other preoccupations—about anything and everything—becomes universal within a culture, it becomes as palpable as the daimon expressed by a person or product. Something "in the air" seems to infect and affect all who breathe it. Germans call this phenomenon the *zeitgeist*—literally the "ghost or spirit of the times." We sense a period's zeitgeist as surely as we sense euphoria or hunger.

The zeitgeist doesn't really exist out there "in the air," of course, any more than daimons actually reside "inside" people and products; it resides within ourselves. Thus, for example, we live in what we aptly call an *information age* because of our preoccupations with computers, cell phones, and fax machines. Information becomes the preferred metaphor for explaining all manner of mysteries as when we liken the brain to a computer that processes information. During earlier, more mechanically minded eras, it was an electrical or hydraulic machine.

Artists and designers unwittingly make the zeitgeist manifest in their works. Like millions of far-flung radios tuned to the same station, the rest of us instinctively resonate harmoniously with any artifact that brings the zeitgeist to mind.

But what does the PT Cruiser have to do with information? It might be prospering primarily from that special kind of information we call entertainment in what Michael Wolf calls the Entertainment Economy. Jeremy Rifkin, president of the Foundation on Economic Trends, suggests in *The Age of Access* that calling this the information age is the "misunderstanding of the 21st century . . . like calling the Industrial Age the Print Press Age." We are approaching an artificial world, he says, when networks and relationships with providers of "just in time" services will supplant ownership of tangible products. We have become, he says, an "experience culture" that would rather pay for the intangible entertainment value a car provides instead of the tangible steel, plastic, and rubber of the car itself. Manufacturers appeal to this frame of mind, says Rifkin, by stressing rollover leases over outright sales.

What makes a transparent iMac seem right just now, when earlier transparent products failed? Perhaps it appeals because, at a time when so much of Silicon Valley's magic lies mysteriously hidden from view, we imagine that we can see beyond the machine's skin to its soul. Perhaps the ways it parses and reorganizes light suggests how a computer parses and reorganizes information.

Perhaps it simply reminds us that transparency is a desirable human trait. Notice how often "transparency" appears in the media today. The world seems increasingly to desire greater transparency in its leaders and institutions. We do not want them mincing words and hiding the truth behind euphemisms. We want the responsible parties to "come clean," whether the issue is malfeasance in office or a product recall. As Gloria Macapagal Arroyo assumed the presidency of the Philippines, displacing a predecessor driven from office for corruption, she stated that one of her highest priorities would be to make government "open and transparent." "Ahaaah!" we psychically drone in unison when we see another example of longed-for transparency, even in the merely symbolic instance of an iMac's transparent shell.

Although information might be front and center in our collective mind, we simultaneously live in several ages. Highways and parking lots tell me that it is also an SUV age. We are preoccupied enough with environmental concerns to call this an *environmental age.* Enough people work out at gyms and jog or ride a bike everyday to call it a *physical fitness era.* And we are still a culture preoccupied with making and owning material things. Thus we still live very much in an industrial age and a consumer age.

The natural evolution and survival of product designs are determined by genetic and environmental factors, as in the case of living organisms. Daimons correspond to the slowly changing genetic forces that shape a product from within. A watch design will not survive a genetic mutation that leaves it looking inaccurate, unreliable, or otherwise unfit for "watch-ness." It also will fail if it differs too much from the norm, just as severe genetic mutations can render an organism unable to survive.

Zeitgeists correspond to the less predictable and more volatile environmental forces that shape products from outside. While daimons seldom change, zeitgeists often do because preoccupations change. Zeitgeists are the engines that drive fads, fashions, trends, and shifting tastes—not daimons. It used to be said that you could tell the state of the economy by observing the hemlines of women's dresses: They went up and down in step with the stock market. Consumers favored black and other somber colors in new cars during the Great Depression of the 1930s. They liked bright pastels during the optimistic 1950s.

During the heady dot-com economy of the 1990s, people preferred the cheerful, even gaudy colors—and forms—of the extremely successful Volkswagen New Beetle to those of the original beetle. Ferdinand Porsche designed the original Volkswagen as a more serious "people's car," during the worst of economic times, for impoverished and disheartened Germans who could not otherwise afford cars. He strove to create a technological tour de force but also a symbol of hope and promise.

The popularity of the Beetle peaked in the United States during another socially turbulent but more optimistic period, the 1960s. So we remember the "flower-people's car" more fondly than, perhaps, the Germans do. It was against this more upbeat background that its nostalgic reincarnation, the New Beetle, was designed—primarily for the American market. Its form came to evoke nothing less than a sense of frivolous joy—just right for the longest period of economic prosperity and social complacency in history.

Transparency eventually will lose its aesthetic potential as it becomes just another meaningless cliché. People who just cannot imagine life without an SUV today may come to loathe them in the future, just as those of an older generation wonder today why the finned fantasies of the 1950s fascinated them so in their youth.

You can discern the spirit of the time in the form of every product. In products we like, the daimon and the zeitgeist they express seem authentic. Certain expectations never change. The expectation of accuracy, for example, would apply to any watch of any era. We expect a watch to keep accurate time, so we are attracted to those which seem precise, regardless of when they were designed. An attractive car of any era will seem mobile, not static. Bright, cheerful colors would have seemed garish and out of place on the depression-era Beetle, whereas they seem all but essential for today's New Beetle. The most attractive watches from earlier eras looked expensive and delicate, like fine jewelry; in today's busy, high-tech atmosphere, we relegate such watches to evenings out and relatively formal coat-and-tie job settings. A robust, relatively mechanical and technical look seems appropriate in most of today's settings.

In a sense the New Beetle and the PT Cruiser are out of sync with their era—except that this one happens to value nostalgia so much for its own sake. Designers can get away with nostalgic looks backward, especially now, but they court disaster when they look so far forward that their designs are "ahead of their time."

Products also express corporate cultures, which have their own zeitgeists. The products of IBM and Apple both reflect today's broader high-tech zeitgeist. But we expect different things from iconoclastic, "Think Different" Apple and "Big Blue" IBM. While the curvy, free form of Apple's original iBook and the strict, linear form of IBM's ThinkPad could coexist successfully in the same marketplace, neither would have done as well by wearing the other's dress.

The designer can determine the proper nature of a product's daimon in a straightforward way with Osgood's semantic differential. Bottling the mental and emotional air of a zeitgeist, however, is more difficult. Nevertheless, artists do seem to have it in their natures to sense and express zeitgeists in their work. George Bernard Shaw, for example, said it was his business to "incarnate the Zeitgeist." The poet Ezra Pound called artists the "antennae of society" because they resonate with nascent social and cultural forces while they are still out there in the ether and undetectable by normal beings. In *Love and Will,* psychiatrist Rollo May likens creative people to some patients in this regard; both tend to sense concerns and crises as much as a decade before "normal" people do. But they are seldom aware of what they are accomplishing. In effect, they are nothing more than unwitting, intuitive vessels. We are all like the proverbial fish that can understand the nature of water only after leaving it; we cannot fully understand a zeitgeist until we also have left it behind. Only then can we see in its icons the zeitgeist's stamp.

The industrial designer, like all artists, has no choice but to proceed intuitively, without the crystallizing perspective of hindsight. The accomplished ones, however, come equipped with artist's antennae that sense what lies just out of range of normal sensibilities. The manufacturer banks on this instinct, often with nothing more than faith.

ICONS

Certain products—such as the Apple iMac, the Palm Pilot, and the Motorola StarTac cell phone—resonate so perfectly with the preoccupations of an era that they become its icons. We know an icon when we see it, but ironically, we often do not see icons coming. And designers seem unable to purposely create one. They just seem to materialize unpredictably from out of thin air.

Religion has been of such central importance to cultures of every age that most icons have been religious objects. The cross, for example, has embodied the distilled essence of Christianity for two millennia. Secular icons of the modern era have emerged largely from the realms of science and technology but, nevertheless, often have inspired nothing less than religious awe as well. Mystic electricity, with invisible, silent, and infinite potential, defined its own era. This invisible and mysterious source of power—quintessentially daimonic in nature—gripped everyone's imagination during the age of electricity. Historian and social critic Henry Adams waxed poetically, not to say religiously, about the 40-foot-tall dynamo at the Great Exposition of 1900 as

> a moral force, much as the early Christians felt the Cross . . . scarcely humming an audible warning to stand a hair's-breadth further for respect of power—while it would not wake the baby lying close against its frame. Before the end, one began to pray to it; inherited instinct taught the natural expression of many before silent

and infinite force. Among the thousand symbols of ultimate energy, the dynamo was not so human as some, but it was the most expressive . . .

The electric car probably rivalled the light bulb as the preeminent icon at the beginning of the twentieth century, when electrics actually outnumbered gasoline-powered cars. Electricity retains a strong grip on our collective imagination to this day. It is the genie we rely on to run our appliances and entertainments, not to mention today's preeminent icon, the computer. Electric trains are the most magical trains, especially the latest magnetic-levitation type that skim along just above the rails, held up by repelling magnetic forces that defy gravity. And when futuristic sci-fi movies show cars, they are always electric cars, running swiftly, silently, and cleanly on batteries that seemingly never need charging.

The factories and industrial processes that made mass production of products possible—and created the need for industrial design—became iconic in their own right. Until late in the 1920s, the 200-year-old Industrial Revolution still struggled to emerge from its handicraft heritage. It finally found its legs and began stretching its stride. Factories began to look and operate more like organic entities, composed of cooperating, interdependent parts—not unlike the products they made. Assembly lines functioned less like means for merely routinizing old-fashioned manual assembly methods. Carpenters in car factories, crafting wooden roof ribs for supporting hand-sewn fabric panels, gave way to powerful, multistory presses that stamped out one-piece steel "turret" tops in single, ear-splitting strokes. Machines in furniture factories cut steel tubing to length and bent it into chairs unlike any seen before.

The icons of production also assumed quasi-religious status. Postcards depicting a city's factories sold alongside cards showing its cathedrals, monuments, and other historic and cultural landmarks. Smokestacks in the artist's renditions belched black smoke that proudly proclaimed the hustle and bustle of the Protestant ethic—inspired by a zeitgeist that did not yet reflect concerns about environmental pollution.

New towns, such as City of Industry in California, were named in its honor. An article in a 1928 edition of *Vanity Fair* called Ford's huge River Rouge,

Michigan, facility—virtually a city in its own right—"an American altar of the God-Objective of Mass Production." The Rouge, as it is still reverently called, not only built cars, it also produced steel, glass, and most of the other materials needed for Henry Ford's Model T's. It was the terminus for Ford's own railroad that brought iron ore and lumber from mines and forests the company owned in northern Michigan.

Products from these factories achieved new levels of sophistication, too, in both form and substance. As designers created products more suited to machine production—and less to hand fabrication—their designs acquired a new machine aesthetic. The appearance of furniture made from chrome-plated steel tubing differed dramatically from that of traditional wood furniture. To avoid the extensive hand labor of upholstery, many chair designs featured much simpler slings of fabric or leather suspended from spare frames. Increasingly, car bodies were stamped from larger and larger pieces of steel that no longer depended on frail, handcrafted wooden frames for support. They had voluptuous crowns that made them stiff enough to support themselves and, in the bargain, to strengthen the bodies they now joined in welded unison. These examples and others that took advantage of industrial technology were genuine instances of the classic maxim that "form should follow function."

The 1930s were an especially auspicious decade for the young field of industrial design to take root and flourish. It turned out to be the most fervently creative decade yet—technically, socially, and artistically. The streamlined airplane, like today's computer, exemplified the leading edge of scientific, technological, and industrial progress—the apex of human achievement.

Art and the Machine, which chronicled industrial design's origins, opened with a photograph of the all-metal Douglas DC-3, the first modern airliner. It epitomized the 1930s so thoroughly that it probably was the period's most significant and influential icon. The plane turned out to be so epochal and seminal in form that its essence can still be seen in all of today's airliners.

The airplane's aerodynamic form influenced the design of other products—not just those which moved through the air such as cars and trains—so pro-

Douglas DC-3

foundly that design historian Donald Bush named the 1930s the "Streamlined Decade." *Streamlines* are the relatively straight and direct paths that streams of air follow from nose to tail of an aerodynamically ideal form. It turns out that a teardrop shape produces the best streamlines and thus requires the least energy to push it through the air: A teardrop-shaped bird can fly farther and faster without tiring; a teardrop-shaped plane can carry more passengers because it needs to carry less fuel; and a teardrop-shaped car gets better gas mileage.

Most cars had teardrop-shaped fenders by the end of the decade. Other products, from chrome-plated teardrop-shaped pencil sharpeners to radios and irons, also appropriated the smooth, gradually curving surfaces of aircraft. Even staid refrigerators assumed the smooth, gentle surfaces of streamlining.

1934 Chrysler Airflow and Union Pacific Streamliner

To say that something was "streamlined" was tantamount to saying that it was fashionably modern and beautiful.

To this day, the term *streamlined* connotes elegance of technology and form. Dictionaries define it as "improved appearance or efficiency," "gracefulness," and "modernity." Referring as well to "effective organization," we even characterize companies with "simplified" procedures and "efficient methods of production" as streamlined organizations. The form of the iMac computer owes as much to the streamlined DC-3 as to the streamlined computer-aided design (CAD) systems that made its free-form shape practical to manufacture.

Little of the functionalist philosophy was evident in 1956 when I arrived in Detroit to seek my first job as a car designer. The automotive capitol of the world had become obsessed with fins—or what the British called "cow's hips"—that simultaneously symbolized technically advanced jet aircraft and spacecraft. Authentically designed fins could have improved the directional stability of cars; they would have made it easier for a driver to hold a straight course in crosswinds or while passing trucks.

Fins on cars like the 1957 DeSoto had little to do with the actual possibility of improving directional stability, however, and everything to do with symbolic power and sex. Cadillac was first to install fins on a mass-produced car. They were inspired by the twin tails of Lockheed's dramatic twin-fuselaged P-38 Lighting fighter of World War II fame. Although quite diminu-

1957 DeSoto

tive in light of those that followed, their unexpected novelty unleashed a storm of controversy.

Controversy, however, unlike universal condemnation, is not necessarily a bad thing. Hardly any artifact becomes significant, aesthetically or historically, without first surviving a gauntlet of controversy. Cadillac's fins did not inspire emulators right away, so they became a unique factor in Cadillac's brand identity—just what General Motors had hoped for. Cadillac's fins grew larger with each successive redesign, but in modest steps calculated not to upset the brand's relatively conservative customers.

It was Chrysler who broke things open in 1956 with fins that ran the entire length of the fender and especially in 1957 when the company introduced fins longer and dramatically higher than anyone had expected. Although the fins reminded viewers of the newest generation of jet fighter, these were not aircraft fins. The airplane now vied with a newer, more dramatic icon, the rocket ship, that tapped the romance of future space travel. Books such as *Across the Space Frontier* and *Conquest of the Moon,* edited by Cornelius Ryan and authored by the likes of Wernher von Braun, the legendary German rocket scientist, ignited the public imagination. The books were laced with tables of factual engineering data pertaining to the design of space ships, orbiting space stations, moon stations, and mission profiles. Coupled with realistic illustrations by outstanding artists such as Chesley Bonestell, they made the prospects for space travel and exploration seem more like imminent science fact than science fiction. New standards of special-effects realism in movies such as George Pal's *Destination Moon* probably had even more to do with the shift of expectations.

Oldsmobile named its Futuramic models, introduce in 1949, the Rocket 88 and Rocket 98. Perched on their hoods were iconic replicas of a rocket remarkably like the one in *Destination Moon.* It symbolized the nation's exuberance and optimism.

Many of the ornamental details on cars of the 1950s had blatant sexual connotations. Rocket-shaped hood ornaments symbolized physical power but, in their overtly phallic connotations, sexual prowess, too. Fins became a phenomenon that every other carmaker then rushed to emulate. Design managers interviewing me seemed interested only in whether my portfolio contained new fin designs that differed enough from Chrysler's in length, height, shape, and angle to set their cars apart enough to improve brand identity. Predictably, my first assignment—and the first thing of mine that went into production—was a fin.

Until then, the streamlined look had a benign demeanor. By their very nature, streamlined shapes did not fight the wind but gave in to it by letting it slip by with as little fuss as possible. The DC-3 was a polite and genteel creature. A typical streamlined car did not stand up to the wind and resist it but rather let it go on its way as quickly as it could. An occasional sharp crease along a fender's crown might slice the air more cleanly but not aggressively. Eventually, a new ferocity and aggressiveness, inspired by warplanes, also emerged. Sharp edges and taught surfaces replaced the benign, gently curving surfaces of prewar cars. Gunsight hood ornaments and machine-gun portals appeared, along with bomblike and rocketlike protrusions.

A DARKER SIDE

By all appearances, the 1950s were largely an optimistic period. Both World War II and the Great Depression lay in the past. Prosperity had returned and brought with it a sense of bigger and better things ahead. Car designers reflected the upbeat mood with such unrestrained enthusiasm and joy that, as far as car design went, it might have been called "the exuberant decade." Cars were extreme in degree, size, and extent. They wore vibrant colors with such names as passion pink and lilac mist (in two- and then three-tone combinations). They wore lavish splashes of bright chrome, stainless steel, and aluminum jewelry that made their predecessors seem somber in comparison.

Yet all periods also have dark sides, even if we don't normally see them in negative light except by hindsight. We prefer to name them optimistically for their desirable and hopeful attributes. Otherwise, elevated concerns for personal safety on highways and sidewalks, even our homes, might lead us to call our time the age of anxiety or the age of violence.

Likewise, the stylized rockets, fins, and bomb-shaped appendages on cars of the 1950s tapped the darker, morose strains of the Cold War era. The iconic rocket perched atop an Oldsmobile's hood could be interpreted as an amulet to allay fears: While it bore a remarkable similarity to the rocket in *Destination Moon,* it also resembled Germany's V2 rocket—the first intercontinental ballistic missile—that rained terror on England during World War II. The fins, bumper "bombs," and taillights that aped rocket exhausts simultaneously symbolized new horizons and the awful possibility of a world-ending nuclear flash just over the horizon.

The Cold War was not the only source of anguish. Like the 1930s, the 1950s brought a contradictory mixture of feverish, optimistic creativity and equally feverish anxiety. It produced the Golden Age of the new medium of television and significant strides in the other arts. It also produced books such as William Whyte's *The Organization Man* and David Riesman's *The Lonely Crowd.* The heat was already being turned up for the sexual, racial, environmental, and political explosions of the 1960s. Sex researcher Alfred Kinsey recently had revealed a concern about male impotence more rampant than imagined, just as men were buying the most powerful cars ever produced, festooned with phallic and military symbols of potency.

So the symbolism of these most expressive of all products had ambivalent connotations. On one hand, they were bombastic expressions of physical and personal potency; on the other, they served as amulets to fend off the terrible potentialities of "the bomb." Because the rocket riding Buick's hood also penetrated a ring, it connoted a phallus. Many cars brandished aggressive, chrome bumper "bombs" that all-male staffs of designers called "Dagmars" in reference to the voluptuous pioneer late night TV celebrity. It really didn't matter who was right about the most basic human drive shaping any psyche or zeitgeist—sex, as Sigmund Freud insisted, or power, as his recalcitrant protégé Alfred Adler contended. The most significant consumer product of all time, at the apex of its evolution, brandished symbols of both sex and power.

I have mentioned only barely what is arguably the dominant American icon of the day, the sport utility vehicle. The SUV's roots date back to the 1930s, at least, when General Motors mounted an all-metal station wagon body on a

1957 Cadillac

pickup truck chassis and called it the Suburban. The SUV's contemporary form emerged some three decades ago when manufacturers began adding passenger amenities to pickup trucks, which enabled them to carry more people with some of the conveniences and comfort found in ordinary cars. Four-wheel drive variants with lots of ground clearance offered a more practical means for getting away to ski slopes and, generally, to more rustic areas for camping and other outdoor activities. Hence, the addition of "sport" to the list of its utilitarian attributes.

The robust, rugged character of these truck-based vehicles had a lot in common, both functionally and visually, with military vehicles. Pioneers of the SUV movement bought surplus World War II Jeeps and Dodge Powerwagons in the late 1940s and 1950s. It was only natural that the makers of Jeeps would capitalize on this by making more civilized vehicles especially for nonmilitary customers—and that a quasimilitary look would carry over into vehicles designed for suburban families.

It is remarkable, however, that SUVs and minivans now account for more than half of all consumer vehicles sold. There is little doubt that people buy minivans for their utilitarian value. While many others buy SUVs for their utility, few actually buy them for off-road sport and recreation. Surveys reveal that people buy them for security. Deep down, they are anxious about something: highway accidents, street crime, carjackings, road rage, or terrorism. The SUV era would seem to coincide with an age of anxiety. Not coincidentally, perhaps, the spurt of SUV sales began at about the time of the Persian Gulf War. And, as SUVs have grown more popular, they have grown overtly more military in appearance. Their rugged looks and mass inspire a bank-vault confidence that all who ride in them will be safe. The high, commanding view of the world they provide the driver and passengers also builds confidence. But the look is not just protective. There is an "in-your-face" pugnacity, too, as though the best defense were an aggressive offense.

Pontiac Rev concept car

A new breed of "crossover" vehicles has brought a friendlier, more docile variety of SUVs based on car platforms instead of truck underpinnings. However, the crossover trend also may yield a more militant variety of family sedans, as presaged by Pontiac's concept car, the Rev.

SELF-FULFILLING PROPHECIES

Zeitgeists influence all forms of creative expression, but all forms of creative expression shape zeitgeists, too. This is especially true when massive audiences are involved—as with TV, movies, and industrial design.

When the cartoonist Chester Gould created Dick Tracy's two-way wrist radio during the 1930s, it sprang from pure, idealistic fantasy. But *it made the real thing inevitable*—only a matter of time—because it tapped some deep and pervasive ideal that resonated with the "science and technology" zeitgeist of the period. Gould did not merely predict wrist radios, as Ezra Pound or Rollo May would rightfully contend; his seminal form was as crucial to the eventual realization of wrist radios as microchip technology. Indeed, Gould's concept may have focused and propelled technological development toward the microchip itself as much as any other factor.

A product such as the Palm Pilot, characterized by seminal form, does not merely respond to a compelling information age zeitgeist; it also brings it into sharper focus for all of us. By directing our thoughts, attention, and motivations in particular ways, the seminal product shapes or reshapes the zeitgeist. Like Dick Tracy's wrist radio, they do not merely presage coming things; they also plant the seeds for them.

From Bauhaus to Broadway

4

I NEVER REALIZED my fantasy of owning a Cord. Even used ones cost too much by the time I reached my 16-year-old rite of passage. So I bought the next best thing, aesthetically, a 1941 Graham Hollywood—which was almost a Cord. The tools and dies for making Cord bodies had ended up in Graham's possession after the Cord ceased production and after brief use by the Hupp Motor Car Company for its 1939 Huppmobile Skylark. From some angles, only a knowledgeable observer could distinguish a Graham from a Cord.

John Tjaarda had designed a new nose for the Skylark, which continued on the Graham. From the windshield back, however, it was virtually identical to the Cord. The zeitgeist of the streamlined decade shows through in the low, streamlined cabin and teardrop-shaped fender forms they shared. The comparison shows how little it takes to significantly alter a product's personality, for the Graham does provoke a different mix of thoughts and feelings than the Cord.

Tjaarda's variation on Buehrig's theme does not seem compromised—just different. The Graham has ample charms of its own. When I show side-by-side photographs of the two cars to people who know nothing of their histories, they are as likely to prefer the Graham's design as the Cord's. I, myself, vacillate between the relatively *rational* Graham and the more *emotional* Cord.

By differentiating the Graham and Cord as *rational* and *emotional,* I have classified them, in effect, by placing them at different places on a continuum that runs the gamut between two extremely different states of mind:

- A *rational,* logical, and thoughtful state of mind we normally associate with science and technology
- An *emotional,* intuitive state of mind we normally associate with the arts

1936 Cord

While Osgood found that any pair of opposing adjectives offered insight into any object or concept when analyzed with the semantic differential, certain pairs seem more appropriate than others and yield more useful information. *Fast-slow* and *active-passive* seem more appropriate as descriptors of cars, for instance, than *sweet-sour.* So why a *rational-emotional* scale? In fact, the *rational-emotional* scale turns out to be one of the most universally relevant scales for characterizing any object or concept. Emotionality has well-known associations with other qualities. An emotional design, like an emotional person, comes across as more active, interesting, and exciting than its rational counterpart—at least until you get to know the person better. It seems somehow busier or more complex. Something more seems to be going on within an emotional design than in a rational one. Bearing in mind that *aesthetic* refers to "sensation," an emotional design has more *aesthetic potential*—more ability to tweak the viewer's senses and get the emotional juices flowing.

We have a natural tendency to differentiate things into relatively emotional and rational categories. The tendency may stem from the fact that the brain consists of two hemispheres, which seem responsible for relatively different emotional and rational mental states. The left hemisphere (which happens to

1941 Graham Hollywood

control the opposite, right side of the body) is apparently the primary seat of rationality. The right hemisphere, which controls the left side of the body, apparently dominates emotional matters.

Other theorists and art historians have used corresponding Apollonian-Dionysian or classical-romantic scales to compare styles and periods of art. (The high-minded Greek god Apollo was associated with clarity, harmony, and restraint; in contrast, Dionysus was associated with ecstasy, frenzy, and irrationality. The rational classical Greek style contrasted similarly with later, more emotionally charged romantic styles.) Furthermore, the *rational-emotional* scale helps to distinguish contributions from the two wellsprings of modern industrial design—Germany and America.

BAUHAUS RATIONALITY

The 1920s were a period of energetic creativity and realization in the arts and industrial technology. Industrial design emerged from an idealized notion that art and technology should be brought together in a synergistic synthesis that could solve most of the world's social and economic problems.

Its products would provide dignity as well as pleasure and utility—at prices that all could afford because of the efficiencies of modern industrial technology.

The unification of art and technology was a natural outgrowth of the aesthetic ideals of the Arts and Crafts movement. But it represented a reaction against the preciousness and exclusivity of products that could be made only in small numbers because they depended on the slow handiwork of skilled artisans. The ideals and inspiration of the movement spread throughout the industrialized Western world, but it was felt nowhere as intensely as in idealistic, rationalistic Germany. There, where many factions of the political spectrum struggled for the allegiance of a nation recently devastated and humiliated by World War I, it promised a practical, harmonious blend of both socialistic and capitalistic ideals.

The spirit of the art and technology movement emanated from a sternly pragmatic and rational philosophy formulated by a revolutionary German school of art and design called the *Bauhaus,* established in 1919. Founders of the Bauhaus expected to foster products with revolutionary new forms that would follow naturally from new manufacturing technologies. The ultimate form of any product would owe little to traditional, largely cultural, influences. The arrangement of a product's elements and their forms would arise strictly from functional and economic necessities of what I later will describe as *agency.* Product form would arise more from the actions of machines than from the hands of artisans. The Bauhaus credo embraced the maxims that "form follows function" and "less is more."

The Bauhaus' rational brand of aesthetics drew strength from perceptual theories of the Gestalt psychologists, who stressed the nervous system's apparent preference for order, simplicity, and symmetry of organization. The Gestalt principles of perceptual organization provided a more rational framework for aesthetics than traditional philosophies. *Gestalt* does not translate precisely from German into English. It means something like "form" or "configuration," but "essence" also comes close. The quintessential Gestalt notion was that perception of the whole (the *gestalt*) differs from the sum of its parts.

The original Gestaltists enjoyed only a brief heyday because they failed to prove their crucial hypothesis that objects of similar shape, color, or size coalesce into visual wholes because of magnetic force fields in the brain. Nevertheless, the principles of visual organization they developed were real enough, regardless of their actual causes, and directly applicable enough to product design to provide sinew for the Bauhaus credo. (Incidentally, the emphases of Gestalt psychology's founders, such as Wolfgang Köhler and Kurt Koffka, differ from those of Gestalt psychology of today. Most contemporary references to Gestalt psychology concern a form of mental therapy associated with Fritz Perls, based on the Gestalt principle of "wholeness," and have nothing directly to do with product form.)

A product designed in accordance with the Bauhaus tradition—governed by unabashed application of science, engineering, and unfettered industrial technology—tends to lie near the rational end of the *emotional-rational* scale.

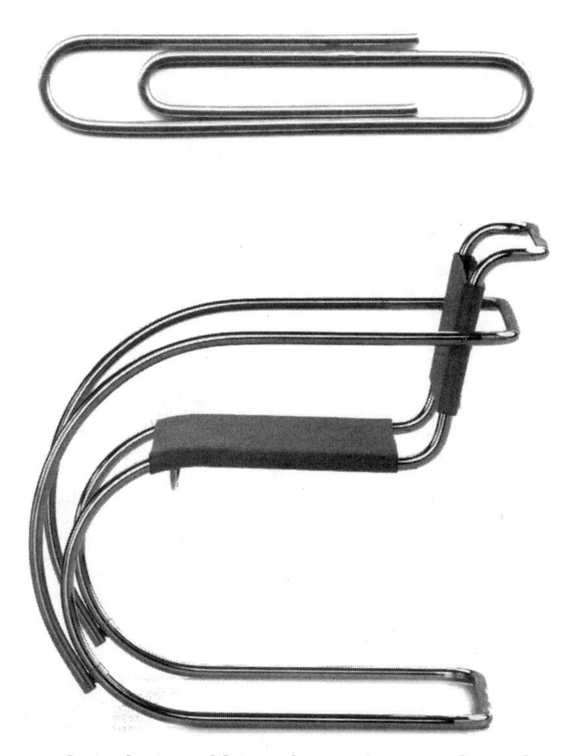

Arm chair designed by Ludwig Mies van der Rohe (1927)

An ordinary paper clip epitomizes the Bauhaus philosophy taken to its limit. Many designers have tried unsuccessfully to improve the elegant minimalism of the paper clip, which is manufactured automatically as we might imagine the 1927 chair designed by Ludwig Mies van der Rohe could be made. A robotlike machine follows a simple program to cut a standard length of wire, bend it in a few places, and drop a finished product into a shipping container faster than the eye can comprehend any step. The beauty and elegance of the paper clip's form—and the chair's—arise from the regularity and simplicity of the processes that produce them. I do not mean to suggest that Bauhaus-inspired designs produce none of the affect and excitement expected of art or more expressive product design; they depend instead on a more intellectual sort of appreciation for their affect. The rational and pragmatic rigor of the Bauhaus manifesto did not preclude artistic fervor. Faculty and students displayed a romantic, virtually religious zeal for artfully applying fruits of the Industrial Revolution to mass-produced products. Their revolutionary zeal, however, lay in bringing technology to art, not vice versa. As a matter of faith, heart-warming aesthetics would follow as a consequence of coldly logical application of technology—with no quarter given for decorative nonsense, affected emotionalism, or arbitrary symbolism. Genuine improvement of agency was the objective: less material, less labor, but more utility.

Nevertheless, such rigorous principles were not enough to stop an occasional lapse into stylistic sleight of hand. A designer might go to a great deal of roundabout trouble to make something look straightforward and practical when, in fact, it was anything but.

Consider again, for example, the circular pattern of dimples used throughout the Audi TT, which has been considered a modern example of the Bauhaus philosophy. While the result suggests that "form follows function," ironically, the dimples are purely decorative. While the Bauhaus manifesto championed a no-nonsense blend of art and technology, the dimples serve only aesthetic ends.

During the Bauhaus' brief history—from 1919 until Hitler closed it in 1933 as a subversive organization—it became, in the words of Elaine Hochman, nothing less than the "crucible of modernism." Although only a legend ever since, it continues to have enough presence and influence to invite attacks from post-modernists, deconstructionists, and Tom Wolfe, who mercilessly satirized it in his book, *From Bauhaus to Our House.* The Bauhaus did fall short of fulfilling its ideals. Its advocates often forced matters to impractical extremes to make them look "practical." And to this day, Bauhaus-inspired products have a preciousness about them that makes them too expensive for most people to enjoy owning them. Still, the tenets of modernism and functionalism espoused by the Bauhaus manifesto prevail in the hearts of today's industrial designers, who regard the Bauhaus as their spiritual touchstone. The building in Dessau that last housed the school was restored and reopened as a design conference center in 1978. It still beckons as industrial design's Mecca.

AMERICAN DRAMA

The upshot of the Bauhaus credo is that paper clips—and often chairs as well—stir emotions largely by appealing to an intellectual passion for order and appreciation of the cleverness, economy, and elegance of the manufacturing processes that shaped them. Consequently a Bauhaus chair often seems cold and aloof to American eyes and hearts, whereas an American chair seems gaudy and frivolous to the German eye and mind. A German brought up on Bauhaus values can derive as much stimulation and pleasure from contemplating a paper clip as a Porsche. For many Americans, though, not even Porsches are exciting enough.

As Sheldon and Martha Cheney noted, the ideals of art blended with technology that shaped the Bauhaus also infused the brand of industrial design that emerged in America: "Industrial design is rightly determined by and geared to industry as it is. The machine is the foundation fact as well as the tool shaping its influence and inspiration." However, Americans produced a different blend of art and technology than did the Bauhaus. Whereas the Bauhaus philosophy stood apart from previous design philosophies specifically by virtue of its new emphasis on the *rational,* technological component, the American school emphasized the chiefly *emotional,* artistic component. American design would lie nearer the more symbolic, dramatic, and emotional end of the scale.

The ascetic aesthetic of the Bauhaus philosophy never took root in America, except in the minds of some industrial designers and a small avant garde especially interested in art, technology, and design—the "early adopters" of the day. On the whole, American tastes have always seemed to gravitate toward more expressive and emotional design. It was fitting, then, that two of the most notable pioneers of American industrial design—Norman Bel Geddes and Henry Dreyfuss—left careers as Broadway stage designers to open their offices in 1927 and 1929, respectively. They set the stage, quite literally, for a more dramatic style that was more expressive, romantic, emotional, and symbolic than the Bauhaus style.

You can see the differences in a comparison of the Cord and Graham. Both cars display considerable elegance, but the relatively restrained and discreet Graham seems more German; the Cord seems brash and boisterous by comparison—decidedly theatrical and American. The difference lies, of course, in the element that sets them apart, their front ends. The Graham's nose resembles the soft, gently curved forms elsewhere in the car. This repetition of form lies at the very heart of visual harmony—just as it does in musical harmony. The overall result is a relatively passive, genteel, or dignified demeanor.

In contrast, the Cord's discordant "coffin nose" parts company with the rest of the car. It was the car's symbolic and artistic tour de force. While its linear, faceted geometry violated the dictates of streamlining, it carried over equally respected art deco motifs popularized by the 1925 Paris Exposition and promoted by outstanding examples of American architecture such as the Empire State Building and the Chrysler Building in New York. It was common to mix streamlined forms with art deco details, but no one had dared to execute such a dramatic contradiction of dominant forms.

Despite Cord designer Buehrig's engineering background and German name, he was steeped in the American strain of the art and technology era that stressed art more than technology. He was as concerned with dramatic, *symbolic* allusions to aeronautical technology as with strict adherence to it. He used the airfoils of the fenders as foils of another sort—to show off the Cord's crown jewel—the brutish hood that transfixed attention on the extraordinary power of the aircraft-derived engine that throbbed beneath. Addition of very unaerodynamic but dramatic external exhaust pipes in 1937 enhanced the allusion. The distinctive shape of the hood also called attention to the car's most unusual technical feature, front-wheel drive, which no other American car had. By pushing the fenders well ahead of the nose (an unaerodynamic throwback to older cars), the wheels thrust forward and visually pulled the car after them—again emphasizing the front-wheel drive. This arrangement also meant that the transmission—which engineers had to locate ahead of the axle—also thrust forward of the nose. Rather than extending the nose to cover it, Buehrig let it show. Continuing to ignore aerodynamic principles, he did not cover it with a streamlined nacelle but, instead, with a more complex cover that amounted to an abstract sculpture of the transmission's machine-like form—again, just in case someone forgot which wheels put the power to the road.

Dutch-born Tjaarda's cleaner Graham probably would have performed better than the Cord aerodynamically. Its smoother, rounder nose reflected the aerodynamic ideal more faithfully than the Cord's blunt, angular nose (when did you last see a bird or a plane with such a blunt nose?). Even the Graham's exposed headlights—adopted as a less costly option than the Cord's retractable lights—do the best they can aerodynamically by resembling the DC-3's engine nacelles.

Had Beuhrig been concerned primarily with functional aerodynamics, his design might have looked more like the Graham's. The Cord might have ended up with a very short hood—wimpy by American standards—or none at all, more like one of today's minivans.

Bel Geddes drew directly from his Broadway experience as he began working with products; he simply designed more attractive and dramatic stages in department store windows for existing products. Soon, however, he also was redesigning the products themselves. He obtained commissions from manufacturers for products that went into production, but Bel Geddes is best known for visionary and inspirational designs such as those published in his 1932 book, *Horizons.* They ran the gamut from elegant teardrop-shaped ocean liners, trains, buses, and cars to giant flying wings, powered by 20 engines, that he imagined carrying 600 passengers. He showed radical house designs and a floating airport that needed only one runway because it could turn into a wind coming from any direction. He is credited with conceiving the modern freeway, with cloverleaf interchanges, for General Motors' exhibit at the 1939 World's Fair.

In the end, many of Bel Geddes' designs seemed too dramatic, too far out, for potential clients. Others were simply too grandiose, fanciful, and expensive to gain funding. So his legacy lay more in shaping the minds of future designers than the actual products he shaped.

The more pragmatic Dreyfuss (who began as Bel Geddes' 17-year-old apprentice) had a more commercially successful career. Dreyfuss broke with Bel Geddes in 1929 after Bel Geddes beaned him between the eyes with an apple core from a theater stage. They had argued about some detail of a set design. On his own, Dreyfuss set about developing long and fruitful relationships with chief executive officers (CEOs) of such companies as American Airlines, AT&T, and John Deere. Before the breakup of AT&T in 1978, Dreyfuss' office had designed virtually every telephone used in the United States. In his book, *Designing for People,* published in 1951, Dreyfuss was an early advocate of ergonomics. Even today, his *HumanScale* guides continue as the primary references for bodily measurements (anthropometrics) used by industrial designers. His office continues after his death, under the direction of his partners.

As abiding as Dreyfuss was of ergonomic and engineering requirements, Broadway nevertheless had imprinted its sense of the dramatic and romantic on his approach to design, as surely as it had on Bel Geddes. Dreyfuss once told me the particularly poignant story of the Mercury, a high-speed New York Central passenger train he designed nearly 40 years earlier. The steam-powered locomotive was one of many commissioned by railroad executives anxious to demonstrate the modernity of their trains in the face of growing competition from fledgling airlines. Old-fashioned underneath newly fashioned streamlined shells, they became icons of the streamlined decade.

The Mercury locomotive was especially simple and elegant; it could have been mistaken for a product of the Bauhaus. Its housing enveloped virtually

The Mercury locomotive designed for the New York Central Railroad by Henry Dreyfuss

everything to create a unified whole that reached almost to the tracks—except for a neatly tailored opening that revealed the three giant driving wheels on each side. Chrome plating made the wheels stand out dramatically from their black surroundings. The complicated assembly of massive connecting rods and levers that drove the wheels—finely machined and polished to a metallic luster—also showed clearly through the opening. However, Dreyfuss felt it needed even more drama.

Drawing on his Broadway experience and reaching even deeper into the American culture, he mounted spotlights out of sight behind the valancelike shell just above the wheels. Continuing his story, but now wistfully, he explained that he had added the lights for the benefit of farm kids along the route who harbored the traditional American wanderlust for faraway places. Piqued by the melancholy wail of the train's whistle each evening, a kid's attention would be drawn to the dramatically staged, steam-enveloped, thrashing and flashing assembly of wheels, rods, and levers.

Lockheed Constellation

By the end of the thirties, a more dramatic successor to the DC-3 was in the works, too. The four-engined Lockheed Constellation, or "Connie," as it is affectionately remembered, rolled out late in 1942. It traded some of the DC-3's rational symmetry for a more emotional, energetic asymmetry. The sinuous, sensuous ogee curve of the fuselage seemed more alive and lively than the tubular DC-3's, reminiscent of a streaking dolphin breaking the water's surface. Beginning with an aggressive, slightly head-down gesture, it arched its back powerfully yet gracefully and delivered a powerful kick of its flukes. Extremely tall landing gear made it visually fly before it left the ground. Inside, the ogee curve presented passengers with a dramatic view of a hill and dale that provided a more interesting stroll through the cabin. Passengers actually crested a hill as they walked from one end of the cabin to the other.

THE POST-WORLD WAR II ERA

After World War II, immigrants from the Bauhaus and their American followers enjoyed a brief architectural heyday with the simple forms of what came to be called the *international style.* Affordable tract homes built in California by the Eichler Company reflected the modernist tradition. Today's modern design aficionados still value them highly, and they have become collectors items. Nevertheless, they and the elegant glass-walled skyscrapers of the international style gave way to more ornate and expressive styles. While some architects have continued throughout the interim to express modernist ideals in commercial and residential projects, such ideals no longer dominate the mainstream.

Things were different inside houses, even quite traditional ones. Refrigerators and stoves did take on a more Bauhaus-inspired look during the 1950s as they shed their curvaceous streamlining for the simple boxlike forms they still have—but not without characteristically ornate handles and other details.

Bent plywood chair designed by Charles Eames (1946)

Cesca chair designed by Marcel Breuer (1928)

Furnishings designed by Charles Eames for the most notable American manufacturer of modern furniture, Herman Miller Furniture Company, had clean but less severe lines than Bauhaus designs. Contrast, for example, Eames' molded plywood and chairs with Marcel Breuer's Cesca chair (designed at the Bauhaus and sold today by the other notable American manufacturer of modern furniture, Knoll International). Eames' plywood surfaces are voluptuous compared with Breuer's planar surfaces. The splayed legs of the Eames design also display more emotion than the vertical, horizontal, and orthogonal frame of the Breuer design.

Even Eames' Aluminum Group chairs—representing probably the most technically sophisticated furniture of the period—were anything but "tubular." Their cast-aluminum frames were organically live pieces of sculpture, contrasting markedly with the colder Bauhaus designs. Incidentally, all these Herman Miller and Knoll pieces verify the value of good design: They continue to be produced and sold today—40, 50, even 75 years after they were designed.

Even the warmest of these designs, however, were too cold for most American tastes. The most popular furniture for those with "progressive" tastes came from Scandinavia. It was characterized by graceful, handcrafted elements of soft, warm teak with "natural" oiled finishes. It was sleek without having the cold look and feel of metal. The Scandinavians had taken the lead ergonomically, so their chairs were unsurpassed for comfort (the Scandinavians had employed lumbar support in chair backs long before Americans designer had heard the term).

We do not have to understand how or why these meanings come to mind in order to appreciate a product's meaning; we only need to experience them. We all have the innate ability to "read" their meaning and to judge the relative emotionality or rationality of two chairs, or two cars, or any other pair of products. Anyone, when asked which of two objects seems relatively more emotional (or rational), can make the determination even when the designs differ only slightly. More remarkable yet, there tends to be significant consensus among judgments. Not everyone will agree with me that the Cord seems more emotional than the Graham, but I expect that most will.

With consensus comes predictability and the promise of an objective way of measuring the empathic expression—and thus the aesthetic properties—of products.

INTEREST AND INFORMATION

The Cord's nose differs from its body so much that it could have come from a different car—even a different era. Its decidedly unstreamlined "coffin nose" strikes a loud, discordant note with the smooth, streamlined forms elsewhere in its body. This stark discontinuity of themes yields the design's most important aesthetic consequence; it makes the Cord more *interesting* and *exciting* than the Graham. I do not mean to disparage the Graham or to say that the Cord looks better—only that it draws and holds the viewer's attention more assuredly and stirs feelings more thoroughly. Indeed, *interesting* is an ambivalent term. Your neighbor proudly shows you his new car and asks, with excited anticipation, "What do you think?" "It's interesting," you answer, not wanting to throw cold water on his enthusiasm by admitting that it doesn't quite suit your tastes. As far as he knows, you might have been agreeing with him. *That a product be interesting is the first condition it must meet in order to be beautiful—or ugly.* At any rate, the Cord's more interesting quality closely corresponds with its more emotional quality. In short, the Cord has more *aesthetic potential* than the Graham. It gives the impression that something more is going on within its form. That something is *information.*

INFORMATION: THE COMMON DENOMINATOR

Your eye offers the best clue to what is going on here. To return to the watches shown on p. 64, the eye spends more time fixed on Watch A than on Watch B at first—even though you might end up liking Watch B more. We can say this, generally, about all designs toward the emotional end of the *rational-emotional* scale—except in cases where the product at the rational end happens also to be especially novel or has some special significance to you as a result of past experience. The Cord draws your eye first because it seems somehow *busier* and more *active* than the Graham. As with other products that fall nearer the emotional end of the scale, it also seem more *complex;* it seems to have *more shape* or *form.* Designers seem to have done more to it, to have *informed* it more. The Cord and any other product at the emotional end embody more *information* than those nearer the rational end.

This comparison of the semantic profiles of two cups with an imaginary ideal cup shows how different levels of information in two designs affect relative aesthetic impressions and evaluations. Cup B embodies more information than cup A. Accordingly, most of its scores lie to the right of cup A's: It seems more interesting, complex, active, dynamic, hot, unusual, and even dirtier.

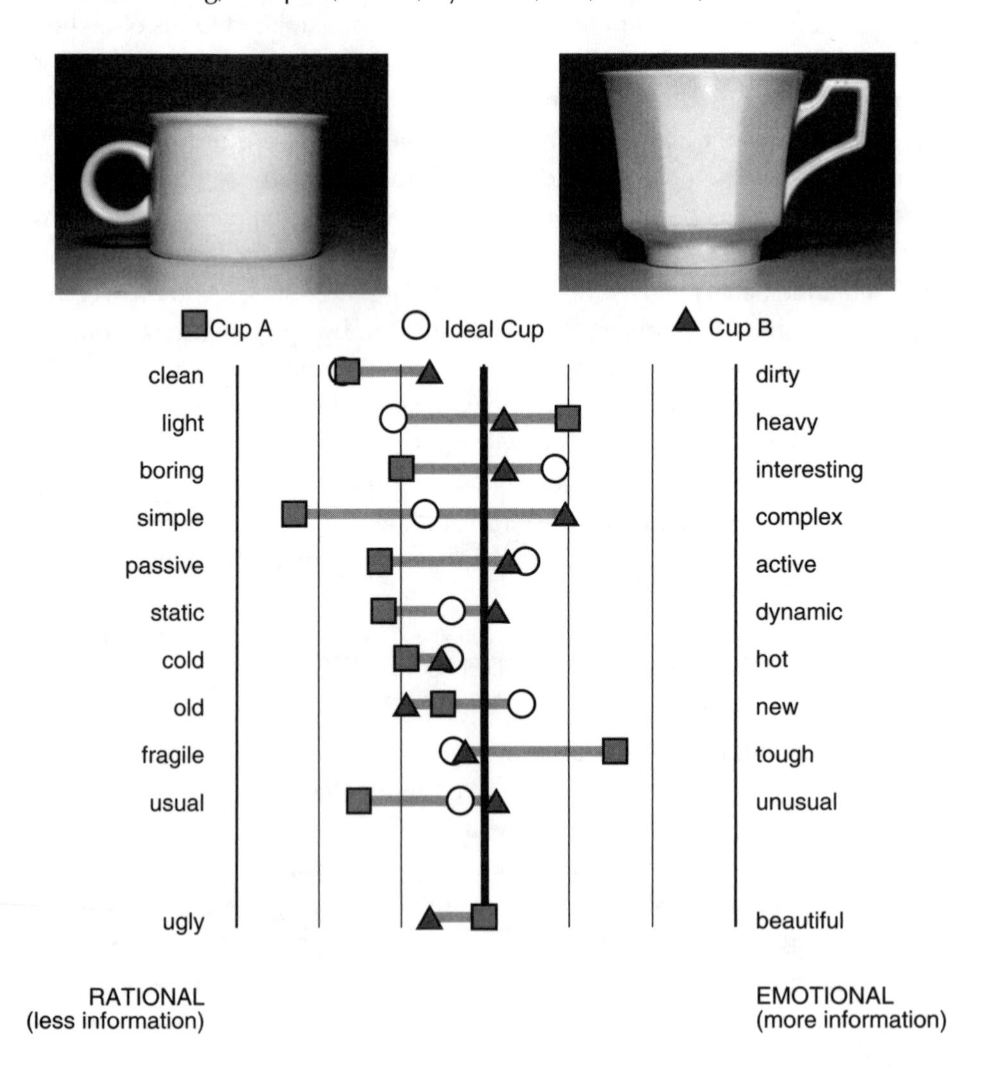

Some judgments are relatively independent of the degree of information. Cup A does not seem heavier and tougher because it has more information. Other form factors dominate. Newness is associated with greater degrees of information, as we will see later, so the old-new scale does indicate that the more modern cup A has slightly more information in this case.

The ugly-beautiful scale is not correlated with level of information. It is strictly evaluative in nature. The judgments suggest that the subjects in this particular survey (all design students) didn't particularly care for either cup, but liked cup A slightly more than cup B. We might also conclude that cup B suffered from excessive information. Indeed, most judgments of the ideal cup lie to the left of the midline. Except for wanting a cup to seem new and interesting (and somewhat active), these subjects would want it to seem relatively rational.

Novelty and significance aside, your eye goes to emotional products first because they embody more of what the eye hungers for: *information.* The amount of information in a thing basically determines our taste for it. Naturally, the eye dwells longer on products with more information because, all things being equal, it takes longer for the nervous system to digest them. A product can contain too much information, however. In such a case, we call it *too complex,* meaning that it contains so much information that it threatens a state of aesthetic indigestion. Isn't this just how you feel when you are faced with an overly long or complicated report—or look at an ugly product (don't forget that ugliness has aesthetic value)? You want to quit by turning away from it in disgust. At other times you may seem to have an insatiable appetite for information. If something seems "too simple," it probably doesn't have enough information to satisfy your appetite.

The fact that your eye goes first to where the bulk of information resides makes perfect sense. The eyes, along with the other sensory receptors, are where the nervous system meets the world. The senses are the system's hunters and gatherers of information. And despite the nervous system's awesome repertoire—as the progenitor and moderator of all feelings, thoughts, memories, and behaviors—information is the only thing that matters. The nervous system seeks out sources of information, acquires information, processes information, and stores information. *And that's all it does.* This means that we can explain any of the nervous system's manifestations—from a designer's inspiration to a consumer's likes and dislikes and everything in between—in terms of what we know about information. So this is where we will turn our attention next.

Products as Media

5

INTERACTIVE PRODUCTS

THE INTERFACE OF an *interactive* or *conversational* product, such as a computer or cell phone, is anything but passive, quiet, or static as it shuttles information back and forth between the product and its user. The interface provides *output* (information the user needs or wants) by means of *displays.* Displays include anything from the complex, changeable array of textual and graphic elements on a computer's screen to the relatively simple face of a watch. The numbers on a phone's buttons also constitute displays. So do the mundane ON and OFF labels on a power switch. Technically, buzzers and other sources of audible information are also called *displays.* The vibrator on your cell phone, which gets your attention without disturbing everyone else in the theater, also qualifies as a display. As previous discussion of empathic expression has shown, the overall look of a product's "body" also qualifies as a display, in the same way that birds and other animals use plumage and markings to attract, arouse, and seduce.

Interactive or conversational products also "listen." They receive *input* (information they need from the user) by means of *controls.* When a user presses a product's power switch, she actually commands the product to wake up and do its thing or get ready to do it. When a computer user presses a key, she provides information that enables the computer to complete one task and move on to the next. A watch requires input too. It remains useless until, at some time, someone gives *it* the time by twisting or pressing some time-setting control.

Actually, most products qualify as interactive devices, even quite simple ones like watches—as long as they have at least one display and one control. This means that any product with a labeled push-button or knob falls within the interactive category. Before a phone user can communicate with someone at the other end of the line, he must first communicate with the phone by observing the numbers on the buttons (output) and pushing a certain sequence of them to specify the other party's number (input). More complex interactive products, such as computers, ATMs, and VCRs, present such mental challenges to users that a new breed of *interface designer* has emerged to facilitate the interactive process at the earliest possible stage of product design. In fact, interface designers usually can improve even simple products if employed early enough in the product development process.

An interactive product demonstrates in the clearest way—as it exchanges information across its interfaces with a user—how products perform as media for conveying information. However, even products that we don't think of as interactive fill roles as communication media. Unlike people, who consciously or unwittingly determine what to say, products serve primarily as intermediaries—media—for conveying what their creators want to say to consumers, users, and the public. What a watch says, for example, is determined totally by its mechanism, which, in turn, has been determined by its designer. Assuming that it has hands for indicating hours, minutes, and seconds—and that the mechanism causes the second hand to jump from one second to the next (rather than smoothly through an infinite number of angles)—it can indicate 12:00 o'clock and 43,199 other discreet moments in time. This amounts to less than the 16 bits of information my computer required to express the number 16 on its screen as I wrote this. A window to indicate A.M. or P.M., which doubles the number of discrete moments to 86,400, only increases the amount of information by 1 bit in the seemingly disproportionate binary math of information science. The point is that it always has this capability, but never more; it cannot indicate 86,401 moments because its designers have not equipped it to do so.

Interactive products, like the computer I write this on, are no less deterministic in principle. Even though it seems more imaginative and original—and capable of infinite variety, for all practical purposes—its mechanism, logic circuits, and software totally determine what it can and does say.

A product's empathic expression, as determined by every aspect of its visible form, is unquestionably also determined by the designer.

THE NATURE OF A MEDIUM

The term *medium* refers to "in-betweenness," as in the case of a T-shirt that is between large and small or water that is between hot and cold or a steak that is between well done and rare. Claude Shannon and Warren Weaver—generally credited with founding the science of information theory in 1947 with publication of *The Mathematical Theory of Communication*—defined the basic communication system. It consists of a sender of information and a receiver of information. The sender, however, cannot inform the receiver directly. In trying to inform you about the subject of this book, I cannot link my brain directly to yours for a straightforward transfer of knowledge. The system requires a third crucial component, a *medium,* such as this book—something *in between,* without which the transfer could not occur. Even sixth-sense science fiction ways of mental transfer always involve wires, electromagnetic waves, "thought waves," or some other medium that creates the mind bridge.

Virtually any substance can serve as a medium—as long as one or more of its attributes can be informed, that is, manipulated, modified, or modulated: ink on paper, paint on canvas, or electromagnetic waves emanating from a radio or TV station. Shannon and Weaver were concerned chiefly with modulating electromagnetic properties of copper wires and radio waves as media for communicating telephone conversations. In the most familiar system, an ordinary conversation between two people, the person speaking at any moment is the sender, and the listener is the receiver. Air serves as the medium. The speaker informs the air by alternately compressing and decompressing it with his or her vocal apparatus. A conversation is a bit more complicated than this, of course, because the speaker (and listener) also use their bodies expressively as well, light being the medium in this case. They might convey as much or more information with their eyes, facial expressions, gestures, and postures as with their voices.

The speaker actually ends up informing the listener to the extent she *reshapes the listener's mind.* The speaker must say something that "changes the listener's mind"—by adding new knowledge or reducing uncertainty about something the listener was not entirely certain of beforehand. If the speaker tells the listener something he already knows or that does not reduce his uncertainty in any way, then the speaker hasn't actually informed the listener. By the same token, the amount of information your watch can provide depends on how uncertain you are about the time when you look at it. If you are very certain because you just looked at your watch a moment ago, you already know, and it bears no information. If you wake up in the middle of the night, however, quite uncertain of the time, it can provide lots of information.

MODES OF VISUAL COMMUNICATION

Products express themselves visually in just three ways. In addition to the body language of *empathic expression* discussed earlier, they employ the letters, words, and numbers of *linguistic expression* and the pictorial symbols of *graphic expression.* Linguistic expression and graphic expression are essential because they pertain to ergonomics; but they are so obvious that they hardly require much explanation here, so I will address them only briefly. I will then continue with consideration of empathic expression, about which much remains to be said.

Linguistic Expression

Products communicate linguistically through labels and other displays incorporating letters, numbers, and words in English or some other natural language. Examples include the ON-OFF label on a power switch, the words of this book, and the numbers on a watch's face. In addition to numbers, a watch usually includes its manufacturer's name (its brand) and perhaps some message pertaining to water or shock resistance. Other products communicate through labels on controls: ON-OFF, HIGH, LOW, CONTRAST, or CHANNEL, for instance. Some, unrelated to controls, issue warnings: DANGER! SHOCK HAZARD!

Government regulations call for labels on food and drug packages that refer to ingredients, nutritional facts, and precautions. Producer-specified labels include instructions for use or preparation and brand promotion. Labels advising users of environmental hazards appear on paints, solvents, cleaning solutions, and garden products. Somewhere on virtually every product, the manufacturer's brand tells everyone who sees the product who ultimately bears responsibility for all it says and does.

The chief advantage of linguistic communication is exactness of meaning in most cases, assuming that the viewer knows the language used. But not in all cases. Many words, especially older ones, have several meanings. Numbers are especially reliable, though, because their meanings are fixed and the same characters used in conjunction with English are used worldwide.

Graphic Expression

A watch's hands qualify as graphic icons to the extent that they resemble pointing fingers. *Icon,* here, does not have quite the same significance as it did in Chapter 3. In this relatively new context of the information age, we have come to know icons as graphic symbols that resemble closely the things they represent. A file folder symbol on a computer display qualifies as an icon because it closely resembles an actual file folder and serves the same purpose. Icons should not be confused with metaphors, like the "desktop" metaphor (the computer's screen) on which file and folder icons can lie and be moved about just as real folders and documents can on a real desktop.

The incremental hash marks that mark the minutes on a watch's face or the increments of a ruler's or thermometer's scale are also graphic elements. The graphic user interface (known simply as a GUI, pronounced "gooey") of a computer's display relies extensively on icons and other graphic symbols. Graphic symbols are more universally understood than linguistic symbols. Users do not have to learn a language, but they do have to learn the association between a symbol and what it refers to. Icons are valuable because they make learning easier.

Nevertheless, the chief advantage graphic expression has over linguistic expression is its universality. Although some associative learning is necessary before a graphic icon or other symbol can be understood, the "language" of graphics is a lot easier to master than any spoken language. Depending on how closely graphic symbols resemble their referents and how familiar people are with those referents, virtually anyone can come to readily understand graphic communication regardless of cultural background. Anyone who has seen a telephone will likely recognize the graphic symbol for a public phone in an airport, for example. And a computer user who belongs to the "office subculture" readily understands that the "document" icon resembling a dog-eared piece of paper can be dragged across the "desktop" of the screen and "dropped" into the "file folder" icon in order to metaphorically "store" it. All these symbols might as well be in a foreign language, however, if someone who has never worked in an office tried to use the computer.

EMPATHIC EXPRESSION

Empathic expression is the closest thing we have to an instinctive, natural, and universal language. People of all cultures learn to express and understand essentially the same vocabulary of facial expressions, gestures, and postures. Because our skill at interpreting empathic messages develops so much earlier than our ability to understand words or pictures, empathic expression affects the viewer from deeper, more innate levels of the object's form than linguistic and graphic symbols can. Words, numbers, and icons might grab our attention first because they lie on a product's surface, but they remain, finally, just superficial. Thus they remain subordinate in importance to the product's underlying form and its empathic expression.

People the world over understand body language and use it expertly without ever having to crack a book or take a course. In the process of learning it, they unwittingly learn to interpret the empathic expressions of products. An infant begins learning empathy, as he learns sympathy, from the intimate relationship with his mother. The infant soon learns to read mother's face, and the emotions it expresses, like a book. Mimicking mother's expressions, the infant establishes mutual communication and the resonant habits of sympathy and empathy.

Standardized emotional expressions and the tendency to mimic them appear so early and seem so innate that those studying them wonder whether they are largely genetic in origin. Evidence suggests that we are born with a set of basic emotions that come with a specific, hardwired pattern of facial muscle contractions for each one. Experiencing a particular emotion automatically causes a specific facial expression. Furthermore, the emotion-expression process is reversible. Flexing the muscles to mimic another person's facial expression arouses the associated emotion. If true, the essentials for sympathy—and empathy—would be in place and functional from the very beginning.

Genetic underpinnings also would explain why empathic expression is understood so universally.

By adulthood, we use skills of both sympathy and empathy instinctively, often unwittingly, but always expertly. Tourists manage to get about in other places with unfamiliar tongues with little more than body language. Products also get along outside their native lands by virtue of empathic expression. People anywhere probably expect a car to seem fast and durable. So, with regard to these characteristics at least, one that did look fast and durable would appeal to people in every market.

With conscientious forethought, manufacturers can use empathic expression to great advantage in designing products that are understood on a global scale. A Japanese truck manufacturer that wanted to design a pickup truck that would appeal to both Japanese and U.S. consumers would find that potential Japanese and American truck buyers have slightly different expectations about truck personas. Both would expect a truck to seem hard and durable, but Americans expect a more aggressive attitude. The manufacturer could design a truck with a hard and durable semantic profile but one that lay between aggressive and peaceful. Different grilles, wheels, and other details would then bias overall appearance toward either an aggressive or peaceful demeanor, depending on whether it would be marketed in America or Japan.

Empathic Metaphors

A watch's hands qualify not only as graphic elements but also as metaphorical hands. Since their significance lies in their relationships to each other and where they point, we could call them *empathic metaphors.* Whether genetic or not, it follows that having developed sympathetic responses to specific facial expressions, an infant would likely experience similar feelings whenever he encountered visual patterns that reminded him of their human prototypes, for example, on the family pet's face or a car, with its eyelike headlights and mouthlike grille.

All products come to resemble living creatures in one way or another, to one extent or another, depending on the shapes and compositions of their "bodies." They can express happiness, anger, bullheadedness, and any other emotion or attitude. We stop well short of actually confusing a car's "face" with a person's face, of course. We aren't fooled quite so easily.

Because stereotypes of Volkswagens, bugs, and beetles closely resemble each other, we can't see a VW Beetle, old or new, without also imagining a beetle or some other real bug. (Real bugs and beetles gained some "VW-ness" at the same time.) This became especially true once the metaphorical names *Beetle* and *Bug* reinforced the visual metaphors.

While these nicknames are endearments today, they weren't originally. They were derived by the public, not the manufacturer; a Volkswagen was just a Volkswagen. However, it looked so unlike any other car that people called it "strange," "ludicrous," and "ugly." The implication that it fit the bug and beetle stereotypes better than the automotive stereotype actually began as an insult. Coined at a time when Americans still ridiculed small cars, they disrespectfully linked VWs with pesky little pests.

Eventually, as Americans came to appreciate the Volkswagen's virtues, they found positive connotations and reasons for endearment in the same metaphors. The hard, beetlelike shell implied sturdiness. As one of the first mass-produced cars available here with independent suspension on all four wheels, it gained a reputation for buglike agility and the ability to crawl over obstacles. Even the invective of ugliness evolved into a term of endearment. Some even dared to call it beautiful as its daimon acquired such desirable attributes.

We don't believe for a moment that these mechanical beetles experience real-beetle feelings or engage in real-beetle behavior, but neither can we quash their compelling emotional and volitional presence, and the illusion that a car can have feelings and the willingness to move. All other products have expressive personalities too, even cold, humdrum refrigerators with little of interest to say.

Linguistics and graphics are optional. A watch needs no numbers to indicate time. Nor, certainly, does it need a brand. However, if the watch has a tangible presence in the real world, it must have an underlying, perceivable form. It needs at least one shape, at least one color, and at least one texture—so it will always express itself empathically.

SPECIAL OPPORTUNITIES, RISKS, AND IMPERATIVES OF EMPATHIC EXPRESSION

The following additional traits entail considerable design opportunities in the use of empathic expression—but also imperatives and hazards.

Aesthetic Significance

Whereas linguistic and graphic forms of expression have primarily ergonomic importance, empathic expression constitutes a product's aesthetic wellspring and the primary source of its daimonic presence. The expressions, postures, and gestures arising from a product's shapes, colors, textures, and *all other aspects of its visible form* shape what and how we *feel* about it more than any other factor. It accounts more for the successes of a Movado watch, the iMac, the New Beetle, and countless other products than any other factor.

Empathic expression also can convey ergonomically important messages. A watch's hands are graphic in nature, but they indicate time empathically by "pointing" in various directions. The shape of a coffeepot's handle can empathically suggest to the user that "I am what you grasp to lift this pot." Practitioners of the *product semantics* design philosophy advocate using a product's form in this manner to instruct the user in the use of the product so fully that she doesn't need an instruction manual.

It Is Irrepressible

Whereas it is possible to design a product that is totally devoid of linguistic and graphic expression, it is impossible to design one without empathic expression. No product—no object of any kind—can maintain visual silence; if it can be seen, it will be heard. As Suzanne Langer observed, every object conveys meaning by the very fact that it has perceivable form. This is so even if no designer or other human agent has shaped it with a particular message in mind. Something as innocuous as a stone lying in a stream, shaped by nothing more intentional than the vagaries of water currents and grinding sand, nevertheless evokes thoughts, feelings, and meanings. It thereby speaks empathically to anyone who sees it. To the extent that each object has a unique form, each makes a unique impression on anyone who perceives it.

The next stone—or the next watch—says something else to the extent that its form differs from the previous one. The designer has no control over *whether* a product speaks empathically, only *what* it says.

A product always says *something* empathically—just as a card player who maintains an expressionless poker face and otherwise behaves in a neutral fashion nevertheless says *something* through his body language. The irrepressible character of empathic expression guarantees that it is impossible to design a watch that tells only time. After stripping away its time-telling numbers and hands and all other instances of linguistic and graphic expression, the residual form of its case, crystal, and strap continues to speak empathically.

It Is Pervasive

Empathic expression is all the more compelling because it is so *pervasive.* It permeates every perceivable element and pore of a product. Every definable aspect of its form affects what it says empathically, including shape, color, texture, and the look of a particular material, be it gold, plastic, or leather. The shapes of a watch's case and strap, for example, must conform to certain functional, practical, and ergonomic requirements. However, the designer still has considerable latitude—in choice of color and material if nothing else—while shaping it.

Ergonomically important linguistic and graphic elements express themselves empathically, too, so they also have aesthetic consequences. The particular font or style of a watch's numbers and the particular shape of its hands convey particular meanings apart from and in addition to their time-telling purposes. Ornately configured hands convey a different impression than simple, linear hands. And Roman numerals make a different impression than more typical Arabic numbers. Of course, no numbers at all makes yet another impression.

In any case, a designer must take care that choices made to enhance empathics do not work against good ergonomics by interfering with messages of linguistic and graphic expression.

It Is Indiscriminate

The most astonishing and hazardous trait of empathic expression pertains to the product's indiscriminate compulsion to answer any question that comes to the viewer's mind. This means that the consumer determines—as much as or more so than the designer or manufacturer—what the product ultimately says. The product addresses the consumer's every concern without hesitation, like the dreaded "loose cannon" who arrives at hasty conclusions without making careful, rational distinctions—and then fires them haphazardly at anyone within earshot.

To the extent that an answer fired off by a product is unacceptable, it loses perceived value in the consumer's eyes. This seems obvious enough if the question pertains to the product's practical purpose. A watch that consistently showed the wrong time would have little or no real value to anyone. Engineers can control this sort of problem, of course, simply by making the works good enough.

Industrial designers can deal with predictable questions associated with identified trends, but no one can anticipate all possible questions, especially trivial or irrelevant ones. A person shopping for a watch implicitly asks each one, "Are you tough enough?" Although a watch's color has nothing to do with its ability to tell time or how tough it is, a stainless steel case will seem tougher than one made of plastic, to the detriment of one made of a tough, high-performance, graphite-fiber-reinforced, but black plastic—except in the case of a consumer who has set her heart and mind on a black watch. Maybe dark colors are in fashion. Or maybe she just wants a watch that won't contrast much with dark clothes.

The irrepressible, pervasive, and indiscriminate nature of empathic expression accounts for why it is impossible to design a watch that tells only time. *A product's empathic expression is, first and foremost, what we notice in its appearance.* It is the most fundamental and predictable determinant of a product's appeal. To attract consumers, its form must *resonate favorably with their expectations and preoccupations.* This means that, above all else, a watch must express those qualities that mark a good timepiece as far as the viewer knows: qualities such as precision, durability, and accuracy.

The empathic expression—the daimon—is so compelling in its effects on consumer emotions and preferences that it can easily override practical, economic, and ergonomic considerations. If a shopper expects accuracy and durability in a new watch, he or she likely will dismiss any candidates that do not *seem* accurate and durable enough—even if, beneath their visible surfaces, they actually are.

VARIETIES OF INFORMATION

Products convey three different kinds of information that can be classified in order of the degree of discretion they permit the designer:

Essential Information

Essential information—such as the time indicated by a watch—permits the designer the least discretion. When a product must embody essential information (not all products do), it always has ergonomic significance. It is therefore crucial to the product's very reason for existence; if a watch did not indicate time, it would be a watch in name only. It does not represent a serious problem with something like an ordinary analog watch because it only needs hands to adequately indicate the time (a digital watch needs numbers, of course). They would not satisfy the more precise measurements required of a track coach's stopwatch; in such a case, numbers and hash marks between them also would be essential. In a more complex product, such as a computer or a power saw, poorly executed essential information could adversely affect efficiency, convenience, comfort, or safety.

Collateral Information

Collateral information, as the name implies, exists *alongside* essential information to supplement it. It is useful, often very important, but never strictly essential; it is always optional. Numbers would qualify as only collateral information on an ordinary analog watch because hands alone would be sufficient under normal circumstances. Stretching a point, even the numbers on a phone's buttons are not strictly essential. A *very* practiced telephone user might tap in a number without looking at the buttons, just as an experienced accountant can use a calculator by touch alone. The rest of us also could make a call without numbers, but not without additional difficulty; we might have to count keys, for example, to be sure which to press. It would take more mental effort and more time. And we would be more prone to make mistakes that would get us wrong numbers. In this case, the numbers are so helpful that they are virtually essential. So you won't see a phone without them. Likewise, collateral menus and icons on a computer's display are not strictly essential (computers didn't always have them), but they make a computer so much easier to use that we regard them as essential enough that all manufacturers now provide them.

The designer has more discretion in shaping sources of collateral information, as long as it does not hinder the effectiveness of any essential information. Collateral information must enhance the product's ergonomic value, usually by increasing speed and accuracy of interpretation. A watch's numbers must be large enough for the user to read them quickly and accurately enough to satisfy the user's needs and expectations. Although the designer can choose among many fonts, some are more readable than others. Numbers and hash marks also must contrast enough with the background to remain readable in the dimmest light expected.

Discretionary Information

The discretion in *discretionary information* refers to the designer's freedom to have the product say anything—about the product itself, the manufacturer, the user, or the owner—without concern for any practical requirements of communication. While not necessary functionally or ergonomically, discretionary information nevertheless can have considerable market value. It is hard to imagine a Rolex watch without a brand and logo on its face or a Mercedes without a three-pointed star on its hood.

Wedding and engagement rings convey socially essential information. Most jewelry, however—like the stunning piece by Arline Fisch shown here—conveys no essential or collateral information. It conveys only discretionary information, chiefly about the person it adorns.

Collar MKC43 by Arline M. Fisch

The exterior design of a car consists almost entirely of discretionary information. By far the majority of a car's surface—its undulations, protuberances, and creases—amounts to discretionary information. Only a few elements (brake lights, turn signals, etc.) provide essential information. A car's interior has much more essential and collateral information, especially within the instrument panel, with correspondingly less room for discretionary information.

At the far extreme from jewelry and automotive exteriors, information products like computers, which convey great amounts of essential and collateral information, leave little to the discretion of the designer. Most of the iMac's discretionary information occurs in the rear of the shell, away from the screen's essential and collateral information.

Chrysler 300 concept car

Regardless of whether a product conveys essential or collateral information, or how much it conveys, it *always conveys discretionary information*—even if only by virtue of something as trivial as its color. Like empathic expression, discretionary information may have little or nothing to do with practical attributes of a product and everything to do with aesthetic attributes. Whenever aesthetic appreciation is the only purpose of the product—as with jewelry—discretionary information rises to the de facto status of essential information.

The manufacturer's brand on a product constitutes discretionary information. So do the specific shapes, colors, and textures of a watch's case and strap. While its hands provide essential information, their particular form and color are matters of discretionary information. Their size, font, and color provide discretionary information, while the numbers provide collateral information.

Where all three kinds of information coexist, discretionary information could be removed without compromising functional and ergonomic objectives. Spain's Grus watches bear no brand identification (discretionary information) but lose none of their time-telling ability as a result. The CEO might argue that the company's name on its products actually qualifies as essential information that helps to ensure the company's survival in the marketplace. For tiny Grus, however, a "no brand" brand could turn out to be among the cleverest of ploys in the branding wars. With no need of copyright, it would remain the "coolest," most exclusive of brands; no competitor would dare risk debranding its watch lest it be confused for a Grus watch and increase Grus' presence rather than its own.

While linguistic and graphic elements unavoidably express themselves empathically—and thus have aesthetic attributes—they serve primarily ergonomic purposes by conveying essential and collateral information. In contrast, discretionary information normally involves *only* empathic expression—and the vast array of opportunities that come with it. *Discretion* does not imply the option of doing nothing—if the product is to succeed in the marketplace. The designer has no choice in deciding *whether* a product says something—but only *what* it says. Discretionary information thus entails the same ominous responsibilities as empathic expression. If the product development team does not exercise its options by determining precisely what the product says, then the product is liable to behave like a loose cannon in the marketplace, saying things the manufacturer wishes it hadn't.

The Inordinate Value of Discretionary Information

While consumers insist that products perform their essential information tasks well, they also seek nonessential, discretionary information in each product as a means for broadcasting personal values, beliefs, and attitudes. As Vance Packard asserted in *The Status Seekers,* products also serve, generally, as symbols of personal status or social aspiration.

The notion that people buy products in order to put on airs does not rest easily with anyone who prides himself as a strictly pragmatic consumer. Practically minded consumers might not consciously buy things primarily to impress others, but neither will they knowingly buy products that suggest inappropriate or unflattering things about themselves. So a person might not wear an unusually expensive-looking watch—out of concern that it would make him seem ostentatious, pretentious, or profligate to others. On the other hand, neither would he purposely wear one that looked so shoddy that it might lead others to think that he lacked the judgment necessary to pick out a suitable watch or was overly stingy.

The value of a bracelet, brooch, necklace, or ring lies entirely in what it says about the one it adorns. In such cases, discretionary information, being the only information, attains the de facto status of essential information; there is no purpose in wearing jewelry except as a means of personal expression. This is largely true for many other products, such as cars and clothing, that do satisfy functional necessities but have little or no essential role as information media in the ordinary sense. This tends to be true whenever competing products fulfill their essential purposes equally well for all practical purposes. Functional performance is less important than price and the appropriateness of the discretionary information in determining the sale.

Even products that serve primarily as information media (TVs, radios, phones, etc.) have important discretionary content. In the early days of personal computers (PCs), looks were not very important. What counted was how well they worked or whether they worked at all. At a time when all prices were high, price did not count for much either. Once the PC business became a market-driven business and PCs became virtual commodities (large volume, lots of manufacturers, comparable performance and prices), discretionary information became more important. Today's computer shopper, able to take performance more for granted, expects good looks too.

Discretionary information increases options for consumers. When every watch tells time as well as the next, regardless of cost, the time may be only a secondary reason for owning one. Indeed, even inexpensive watches have become so reliable that some people own several "fashion" watches. Each one has a different look for a different occasion (business, sports, dining out, etc.). They continue to serve as timepieces, but they serve primarily as jewelry.

The Expression-Information Matrix

We can summarize the relationships among the three modes of expression and the three classes of information with a 3 × 3 matrix. Using an ordinary analog watch as an example, each cell suggests the purposes and benefits associated with each combination.

- The empty essential/linguistic cell reminds us that numbers are not essential for telling time with an analog watch. If we considered a digital watch, this cell would contain its digits and a colon separating hours and minutes.
- The hands in the essential/graphic cell provide the only truly essential information. We know they are essential because if we removed them (or

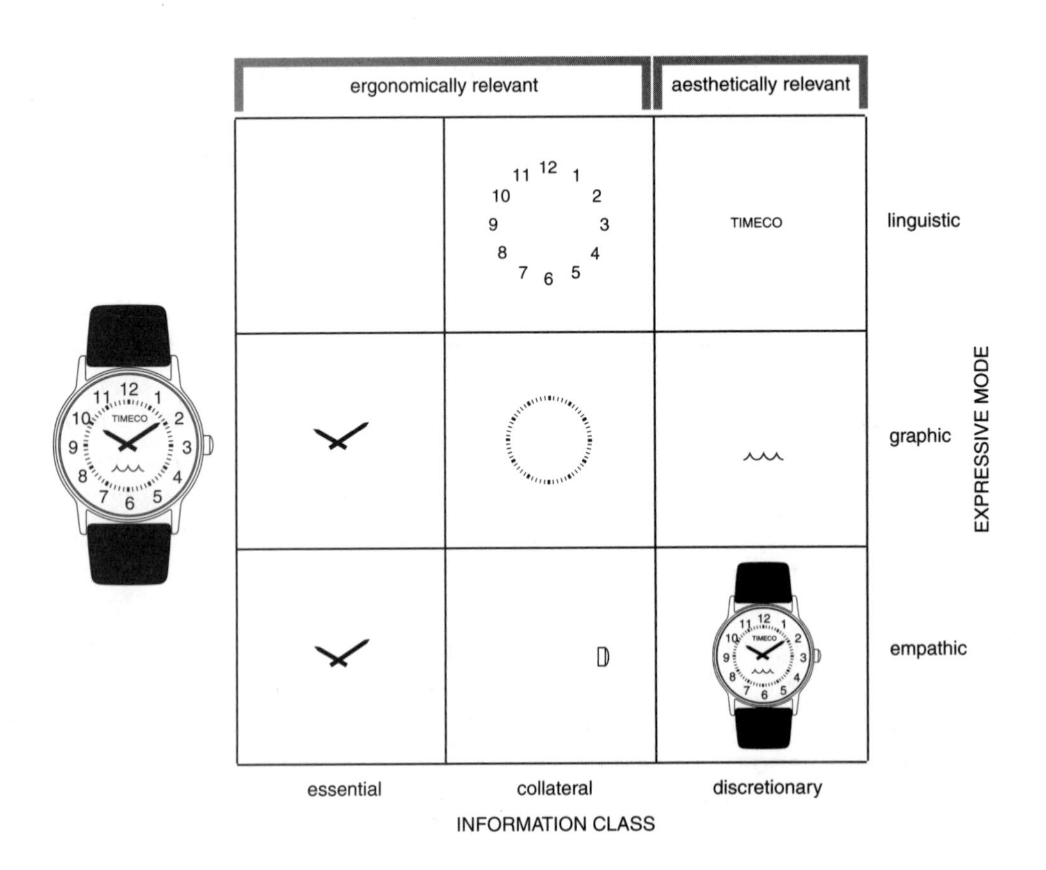

the numbers of a digital model), the watch would lose its ability to indicate time.

- The hands also appear in the essential/empathic cell because they indicate each instant of time with a distinct postural relationship.
- The collateral information column contains the numbers in the linguistic cell and hash marks in the graphic cell. The stem qualifies for the empathic cell because it calls attention to where and how to adjust the watch and the user imagines grasping it.
- All cells of the aesthetically relevant discretionary information column are filled. All visual elements, regardless of their primary purpose, have aesthetic consequences—while none are essential with regard to practicality or ergonomics. The empathic cell is most significant because it includes the entire watch, which indicates that every perceivable nuance—including the hands, numbers, hash marks, stem, brand, and graphic symbol—communicates something in addition to whatever its ostensible message might be. The manufacturer's brand is linguistic, but optional. The equally optional graphic wavelike symbol suggests water resistance. Decisions made with respect to every other cell in the matrix also affect this cell and, thus, aesthetics.

More generally, we can say the following:

- Some products convey no essential or collateral information.
- No product can convey only essential or collateral information.
- All products convey discretionary information—by virtue of empathic expression, if nothing else—regardless of how much essential and collateral information they also convey.

A MULTICHANNEL MEDIUM

In considering the several audiences a product must appeal to, the two most important are *users* and *consumers.* Everyone else exposed to the product comprises a larger and broader *public* audience. Members of the public presumably are not in the market for such a product at the moment. The press and other media constitute an important part of this audience. Benefits follow when this public audience also likes the product and boosts it through word of mouth.

In all, there are at least six channels of visual expression through which a product broadcasts to its audiences. There are two with primarily ergonomic significance (affecting product *performance*):

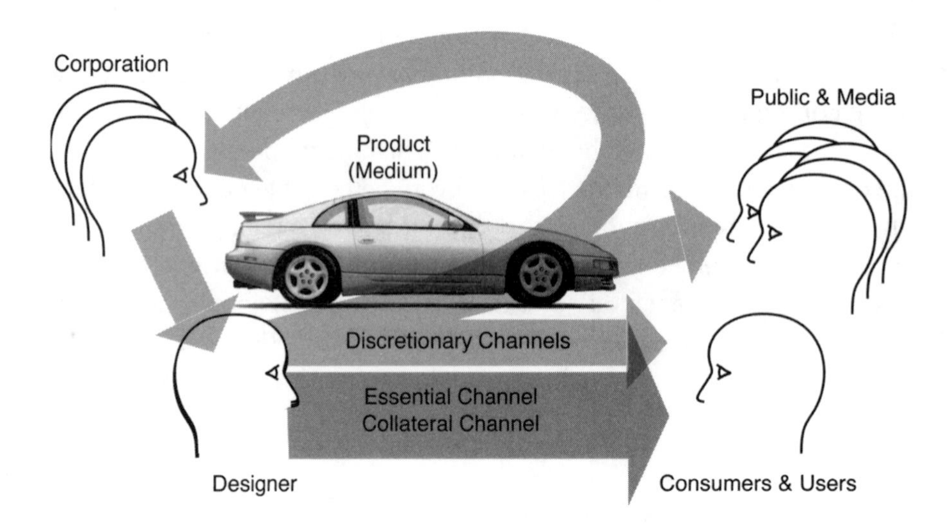

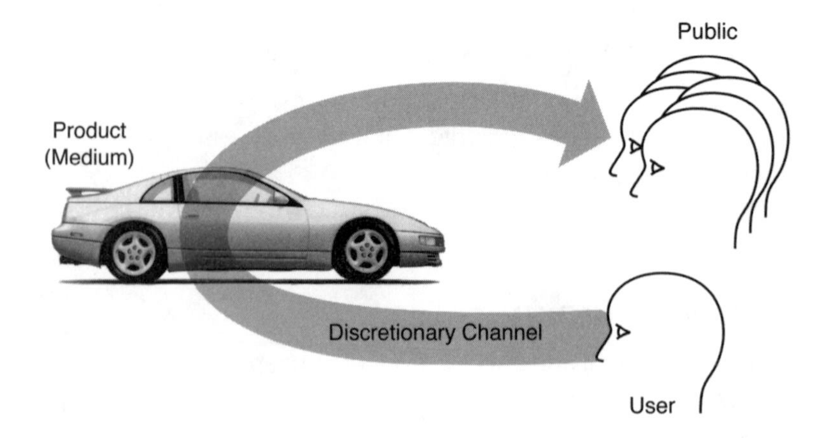

Channels of communication

- Essential information to *users*
- Collateral information to *users*

And there are four with primarily aesthetic significance (affecting product *semblance* and corporate *image*):

- Discretionary information to *consumers*
- Discretionary information to the *public*
- Discretionary information from *users* to *consumers* and the *public*
- Discretionary information *back to employees*

The last channel on the list, which suggests that the manufacturer's employees constitute a fourth audience for discretionary information, is less obvious than the rest. So it never shows up in the official product requirements document or strategic design plan. It ought to because the potential of good design for enhancing a company's fortunes through this channel is significant. Employees who are proud of their products are more satisfied with their jobs, more loyal, and perform better—with the compounded result that they make even better products. This has an upward-spiraling effect on morale, product quality, and what might be called the company's zeitgeist—as well as the bottom line.

Form and Information

6

IF THE VAST array of products that occupy our awareness each day constitutes a mass-information or communications medium, as I contend, then we can study product form in the context of information theory. I refer to products as mass media quite literally. Each product constitutes a crucial component of a communications system. It is the component—literally *in between* the broadcasters of information (manufacturers and designers) and several audiences—that enables the whole process. Designers shape—literally *in-form*—the product; the product, in turn, informs each member of each audience. In fact, information provides an appropriate context for understanding all aspects of product design, from the creative and expressive urges of the designer to the formation of consumer preferences, by considering how the nervous system acquires, processes, and stores information. Information, after all, is the nervous system's only resource and commodity.

MEANINGS OF *INFORMATION*

Information is another of Uwe Poerksen's "plastic" words: It has been cast and recast many times as it has passed back and forth between common, everyday usage and technical or scientific jargon. In the process, it has accrued a confusing array of psychological, mathematical, physical, and vernacular meanings. The foremost definition in my dictionary is "knowledge." It is also defined as "a collection of facts," "communication," "a lack of redundancy," and the "bits and bytes passing in and out of computers." In each of these contexts, information is a "collection of facts or data," something stored in the mind, "the sum of what has been perceived, discovered, or inferred" over a lifetime. In this context, information is a *commodity* that can be acquired, shared, stored, or lost.

However, *information* also implies a *process.* The word stems from the Latin verb *informare,* "to fashion," "to shape," or "to give form to" something. Without realizing it, we commonly mix these notions of information as commodity and process. We say that a teacher *informs* her students and that an *informed* person is a *learned* or *knowledgeable* person. However, we do not refer to designers or teachers as *informers*—which they literally are—because this use of the word now entails a negative connotation, as someone who snitches on others. It is even accurate to think of information as literally shaping the brain. Recent studies have confirmed that learning increases the length and number of threadlike dendrites in the brain that transport impulses among nerves, as well as the number of their interconnections; the brain actually gains matter and complexity.

We can see how this applies to product design if we imagine the process of making a pot, one of the simplest of all products. The potter begins with an *amorphous* (formless) lump of clay to which he or she *adds information* by *informing* the lump until it looks like a pot. We have a problem of terminology right away, though, because an amorphous lump is anything but devoid of information. The real problem is that it has so much *meaningless* information that we can't find a recognizable pattern of a pot or anything else in it; we can't *make sense* of it. In other words, we can't name the shape anything else but a *lump.* When the potter begins to shape the lump, she actually *removes* information by smoothing out the variations in the surface. And this brings us to the most basic definition of information. In the final analysis, a perceivable *variation, difference,* or *change* of something underlies and unites all notions of information.

VARIANCE: CONTRAST AND NOVELTY

The nervous system is organized to register variance. In fact, Charles Babbage, the nineteenth-century creator of the predecessor to modern digital computers, coincidentally called his information processor a "difference engine." At one extreme, the information in a product's form involves differences of lines, surfaces, colors, textures, or any other discernible attribute of its form. At the other, information as knowledge involves a change of mind. Furthermore, there are only two basic kinds of variation to be concerned with—*contrast* and *novelty.*

Dan Berlyne, professor of psychology at the University of Toronto and an early leader in the revitalization of psychological aesthetics, called contrast and novelty *collative variables* because they depend on comparisons, or collations, of two or more things or properties. For example, the information in these words depends on a comparison of the colors of the ink and paper. To the extent that they differ, they have contrast and, therefore, information. If the ink on this page were the same white color as the paper, a comparison would reveal no difference and, therefore, no information. Indeed, you could not read these words in that case. Novelty involves comparison of what you expect to happen with what actually happens. If you expect the new TV set that just arrived to have a rectangular screen but it has a triangular screen instead, you would regard it as a very novel TV set.

Berlyne listed other collative variables that amount to special cases of contrast or novelty: Complexity, ambiguity, puzzlingness, and incongruity are all instances of contrast; surprise is an instance of novelty. In summary:

- *Contrast* involves comparison of two *objective* and *measurable* attributes (such as directions or curvatures of a line or surface, colors, or textures) that differ from each other. I call contrast *objective information.*
- *Novelty* involves comparison of a real object with a *subjective* mental model—a *stereotype* that amounts to a generalization of all similar objects seen in the past. It is most influenced by those seen most often and most recently. Because the stereotype cannot be observed or measured directly—but only indirectly by such means as the semantic differential—I call novelty *subjective information.*

CONTRAST

The contrast of the black printing on this white page qualifies it as information; consequently, your eyes sense it, and the brain begins chewing on it, trying to make sense of it. This would have been much more difficult, if not impossible, if the printer had used white ink. Assuming that the ink would have had a different gloss, you might have been able to read it by carefully turning the page so that the light fell in just the right way for you to distinguish the different glosses of page and ink.

While the page contains information, it is not necessarily informative. To be informative, the meaning of my words must seem novel to you. If it doesn't—if you already know what I am trying to communicate *and believe it*—my words will only bore you. If you already know it but don't believe it, then my words might irritate you. Only if you don't already know it—and my words *change your mind* by modifying some concept or belief—will I have succeeded in informing you.

In the latter case, this book will have completed part of its mission as an information medium. The printer and I will have informed the page by arranging the ink just so on the contrasting paper, and the page, in turn, will have informed you by rearranging the neural stuff in your head (literally *informing* or *reshaping* it) so that you and I now hold at least one belief in common.

I can call contrast *objective information* because it involves comparison of objectively measurable properties that are available to the senses of more than one person at the same time. You and I perceive the same contrast of this printing and paper. We can confirm this by measuring the reflected light from both with a light meter and dividing one measurement by the other to obtain a ratio that corresponds to a standardized measure of what has been defined as *contrast.* By the same token, we can measure the curvature of two curves and determine their contrast. We still won't know whether what you see subjectively as red is what I see, but at least they correspond to the same thing we call *red* in the objective world as measured by a spectrometer.

More than Black and White

The two watches that follow illustrate the most familiar kind of contrast, that of *value* or *brightness* contrast. The watch on the right—with numbers and hands that contrast more with their background than those of the other watch—has more information. The implication of their difference is not obvious under appropriate lighting conditions at normal reading distance; for all practical purposes, you can glean the information equally well from either watch, but not if you move farther away or progressively dim the room lights. Under such circumstance, the one with less contrast will become indiscernible before the other.

Low Contrast (Less Information) High Contrast (More Information)

Value contrast

It could be that your attention is drawn more to the one with least contrast. This would seem to contradict the notion that attention always goes to the greatest concentration of information. The one with least contrast may be drawing your attention because its pale contrast is more *novel* than the other.

The two watches shown here illustrate shape contrast. The watch on the right grabs more attention than the one on the left because its shape varies more. It

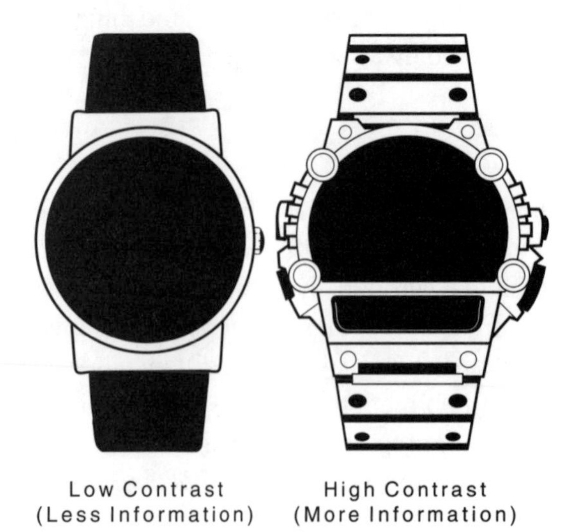

Shape contrast

gives the mind more information to chew on; it seems more complex (*complex* and *complicated* are words we use for "lots of information"). It takes more effort to memorize it and remember it, too; it would be more difficult to draw an accurate picture of it from memory. Finally, the greater mathematical complexity of the image on the right required more of my time to construct it and more bits of my computer's memory to store it.

The Eyes Know

You do not need a computer to tell you which of two objects has the most information. Your eyes tell you; they dwell longer and more frequently where information is greatest. Like all your senses, your eyes serve not only as gateways to your nervous system but also as active hunters and gatherers of food for thought. They have a predisposition for the most fruitful hunting grounds, which always lie in variegated regions of the environment, never in the "plains." Given an environment lush with variety, they go instinctively to the region of most pronounced variety. Given the choice of straight and bent lines, they will go to the bent one every time, the most *interesting* one, the one with the most information.

By the same token, the undulations of a sports car's surface, with numerous changes of direction and curvature, is more eye-catching and interesting—and exciting—than the sedate sedan with its gradual changes. It tweaks the nervous system more than the simple one and gets the emotional juices flowing more turbulently.

The eye and brain need nothing more than regions of extreme change—where information is concentrated most densely—to recognize the Volks-

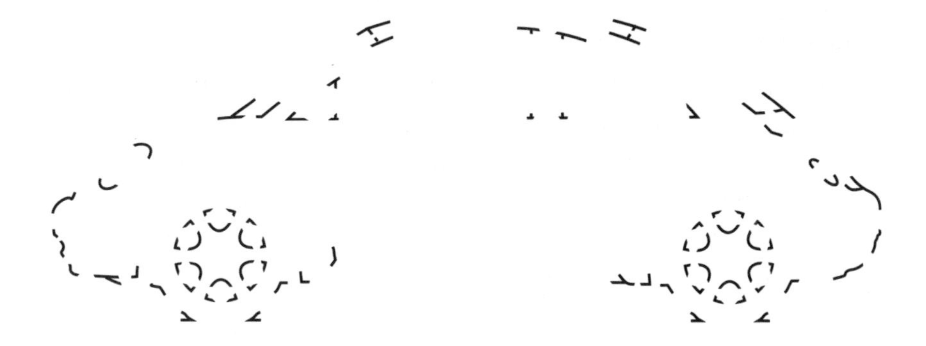

wagen in this illustration despite its sparseness. The mind easily "connects the dots" with imaginary segments of minimal information in order to complete the picture. Curved areas on the car's surface are illustrated with more line segments here because they contain more variability and thus more information that must be reproduced for the illustration to be recognizable.

Many more concentrations of information, in the form of contrast, appear in an actual object: contrasting colors of object and background and contrasting colors of elements within the object (tires and body, for example). Other instances enhance perception of the object's properties: contrasting values of light and dark ("shading") due to the angle of incident light rays and contrasting colors of reflected environmental colors (of sky and ground, for instance) that characterize shiny surfaces.

Being so readily recognizable with so few cues suggests that the New Beetle's design has a clearly defined image that viewers readily recognize and are not likely to confuse with other cars. Such clarity is important to brand recognition and should be a part of any design strategy.

While vision dominates the search for information (we acquire most of our information by far through vision), the eyes are not lonely hunters; other sensory receptors hunt alongside them and behave in essentially the same fashion. With eyes closed, your fingertips can find no nourishment as they roam the smooth, featureless surface of this page—until they find an edge, the first pronounced change of surface direction. Instinctively, they explore the edge until they find a corner, a region of even greater variation and fascination where the surface changes in two directions.

The Choosy Nervous System

As a practical matter, the nervous system has to ignore some sources of information because of its limited information-processing capacity. The environment is so variable that it could easily overwhelm any nervous system that tried to process all the available information. As a matter of survival, the nervous system has to distinguish between trivial information, which poses no threat, and that which is more likely to be important to well-being. Otherwise, trivial matters might preoccupy us so completely that potentially life-threatening events would go unnoticed, to our peril.

So your nervous system is more sensitive to some kinds of variation than to others. It does not register a so-called first-order phenomenon such as velocity, which is a *change of place.* However, you are aware of the second-order phenomenon of acceleration (or deceleration), which is a *change of change of place.* This difference shows up when you take a plane trip. You are aware of the plane's acceleration on takeoff and even during the slight deceleration as the pilot backs off on the throttle to begin descent, which are both second-order effects. En route, however, you have no sensation of how fast the plane is traveling or that it is moving at all as long as it flies at constant speed—especially with your eyes closed and no visual cues of movement with respect to the ground or clouds outside the plane. You don't know whether you are flying at 600 or 6000 miles per hour or, indeed, whether you are still parked at the gate.

This "flaw" in the nervous system serves, in effect, as an *importance filter.* It enhances survival by turning your attention toward events representing the greatest potential danger while ignoring unlikely threats. It does not bother you with an awareness of the earth's constant spin and nearly circular motion around the sun, for instance, or the fact that you are hurtling through the universe at incredible but constant velocity. Instead, it rivets your attention on second-order effects (such as acceleration) that, in the physical world, have more serious implications than first-order effects. As the wag once noted, "It's not the fall from a ten-story building that kills you; it's the sudden stop at the end."

It is important for changes of velocity (acceleration or deceleration) to get your attention because they signal potentially harmful transfers of energy from the environment to your body. The minor deceleration you sense as the plane slows for landing represents nothing to worry about because the amount of energy transferred through your seat belt is well below the harmful threshold. However, the major deceleration you sense as you hit an unyielding sidewalk at the end of a fall is definitely worth worrying about because it represents an enormous transfer of energy, even if you haven't fallen from a ten-story building. A fall while merely walking on the sidewalk probably won't kill you, but it can transfer enough energy to injure you.

The nervous system is wary of all changes, not just changes of velocity, because they all signal concentrations of energy that portend danger. You might get away with a few bruises if you fall on the flat, unchanging surface of the sidewalk. But you will almost certainly break something if you fall against a curb, where the concrete suddenly changes from horizontal to vertical orientation. So the dangerous edge of the curb stands out in your mind more clearly than the relatively safe flatness of the sidewalk. In the final analysis, then, the nervous system isn't merely on the lookout for concentrations of information, but the concentrations of potentially dangerous energy they represent.

Looking through the same importance filter, a curve's radius is less important to the eye than changes of radius. This is why the sphere is the closest thing in the three-dimensional world to an information-free object. No object could be totally devoid of information, of course. Without information to tweak the nervous system, we could not perceive it. A sphere has no concentrations of information due to changes of curvature, so the eye roams freely over its surface, hanging up nowhere. Except for the constantly changing direction of its surface, the sphere otherwise maintains constancy in every other way we might choose to analyze it. Its radius is the same at every point, and its curvature is the same in every direction. By definition, its surface constitutes an infinite number of points, all lying the same distance from its center. Its profile is always circular, regardless of how you turn it. Slicing straight through it always reveals a circular cross section. It is the epitome of continuity, minimal variance, and blandness. However, the slightest dimple, pimple, or crease arrests attention and holds the eye or fingertip.

This points up why even slight discontinuities in a product's surface stick out like sore thumbs. Even though the information gain of just 2 or 3 bits seems trivial, it is concentrated so densely, at an infinitesimally small point, that it delivers a relative wallop to the nervous system. Such a minor discontinuity represents no real threat, of course, but the nervous system can't help itself. It is organized so that it must make a big deal out of it. Better safe than sorry! We benefit from our tendency to be Nervous Nellies: Our ability to be thrilled by a beautiful object depends on our ability to be needlessly frightened.

Uncertainty and Reduction of Uncertainty

The most recent incarnation of *information*—to be used and confused alongside all previous incarnations—concerns the bits, bytes, and databases created, manipulated, and stored in computers. Conceived as something transportable over phone lines and other communication links, it reinforces the notion of information as "stuff." Claude Shannon called *information* simply "that which reduces uncertainty." He referred to more than the psychological notion of uncertainty. He was concerned with statistical notions of probability implied by such questions as, "What are the chances that heads (or tails) will come up if I flip a coin?" The answer, of course, is, "One chance out of two possibilities," a statistical probability of 50 percent that corresponds to one bit of information.

Your car's oil-pressure warning light provides the same one bit of information. It is the simplest of all information displays, capable of indicating just two possible states: As long as the light remains off, it indicates that the oil pressure is high enough to maintain the engine's health; it comes on only to indicate that pressure has dropped to a level too low to prevent engine damage. It cannot tell the driver what the specific pressure is; it would need more bits of information to do that.

Shannon was concerned with less trivial issues than oil-pressure warning lights, such as the probability that this entire paragraph could be sent over a phone line to arrive intact at the other end—instead of one of the virtually infinite possible permutations due to the static called *noise.* Lengthening the paragraph would increase the amount of information it contained—in terms of both bits (computers normally use 8 bits, or 1 byte, for each character) and greater uncertainty about whether it would get through flawlessly. Thus, increasing information *increases* uncertainty. But Shannon said that information *decreases* uncertainty. How can we reconcile this apparent contradiction? Again, we can blame Poerksen's plastic word phenomenon for the confusion. The apparent contradiction actually corresponds to a neat, if ironic, symmetry of ancient and modern notions of information.

The uncertainty that gets reduced is the old, familiar kind inside your head, as when you aren't sure how long something is. Consider measuring the width

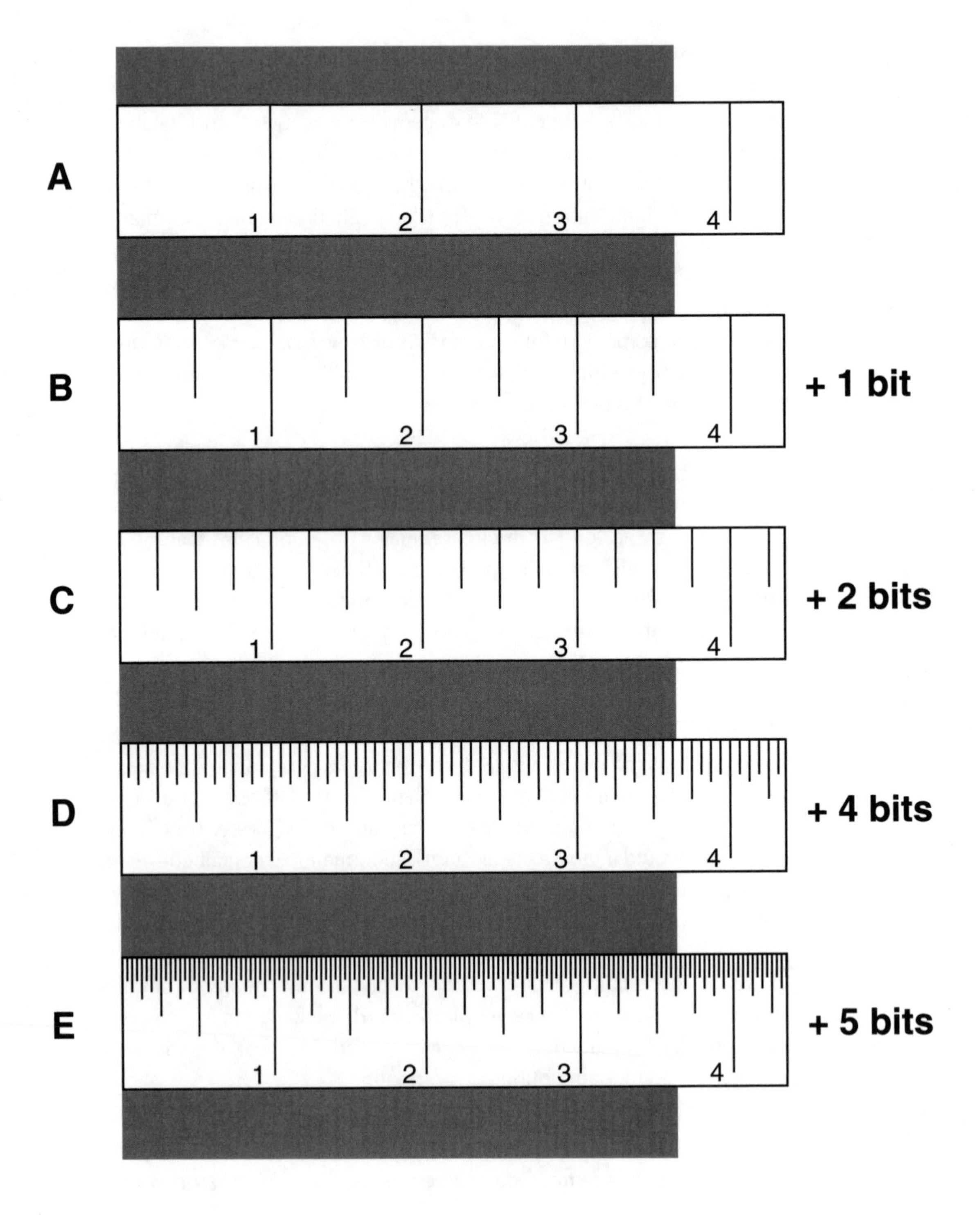
A
1
2
3
4
B
+ 1 bit
C
+ 2 bits
D
+ 4 bits
E
+ 5 bits

of the shaded rectangle of this illustration with an ordinary ruler, one of the oldest and simplest sources of essential information.

Ruler A, divided into twelve 1-inch increments, cannot reduce the uncertainty about the dimension in question entirely because no hash mark coincides exactly with the rectangle's edge. I am reduced to guessing the rectangle's width, but I am quite certain it lies between 3½ and 3¾ inches. If I want to reduce my certainty further I need a ruler with hash marks that are closer together. Ruler B, with hash marks that are just ½-inch apart, does this. However, it presents a higher level of uncertainty than does Ruler A because there is only one possibility in 24 (instead of 12) that the object being measured will fall between any two hash marks. This is equivalent to increasing the amount of information the ruler can convey by one bit. The interesting thing is that my own uncertainty decreases as the ruler's uncertainty increases.

Actually, Ruler B has more than a one-bit advantage over Ruler A because it also has more collateral information. The numbers on both constitute collateral information. Like all collateral information, numbers aren't strictly essential to a ruler's purpose but they lead to quicker, more accurate measurements. The additional hash marks of Ruler B provide more collateral information, as well as essential information, because they have a different length than the 1-inch marks. Again, this contrast of lengths is not essential. It is not even very important in this case. Contrasting hash marks do become more important as their numbers mount, however, as in Ruler E,

Each time a ruler is subdivided by doubling the number of hash marks, as shown in Rulers C, D, and E, it gains another bit of information and I grow more certain of exactly how wide the rectangle is—up to a point. The marks come so close together at more than about 32 increments per inch that my eyes have trouble distinguishing them. Once the marks visually fuse, I am confused and my uncertainty begins to rise.

Like any computer, a nervous system has limited ability to acquire and process information. Ruler makers acknowledge this ergonomic limit by making rulers no more with 32 or 64 increments per inch.

This limitation of the nervous system also means that laboratory timepieces that slice time into millionths or billionths of a second cannot be read "on the fly." They can record only elapsed time by stopping and freezing ongoing time in its tracks. A track coach cannot use a stopwatch to tell time as it passes, a hundredth of a second at a time; his nervous system cannot keep up. In the time taken to mentally register one increment—if he could at all—the watch would have raced many increments ahead. A digital stopwatch that displays fractional seconds demonstrates the problem nicely. Forget the hundredth digit; it sequences so quickly that it appears to always display a flickering 8. Although you can distinguish different digits as they whiz by in the tenths-of-a-second place, they are of no practical use until you stop the watch. Full seconds lie at the practical threshold of perceptual abilities. So, second-long increments are about as finely as a watch can slice time if we are to read it "on the fly." Not coincidentally, this is as fine as ordinary watches slice it. Most of the time, seconds are too small to be useful. You would never tell someone, "Meet me at 1:15 and 20 seconds." Still, such watches are nice to have for counting down the time until a rocket's blastoff.

Budgeting Information

Given the nervous system's finite capacity for acquiring and processing information, information must be budgeted in ways that reconcile ergonomic and aesthetic needs in light of established priorities. Designers must give highest priority to ergonomically important essential and collateral information. Any of the user's information-processing capacity that remains can then be allocated to aesthetically important discretionary information.

The various keypad designs for a hypothetical phone in the accompanying illustration show ways of budgeting information. Design A has so much discretionary information (contrasting button sizes and colors) that it would adversely affect its usability. The unorthodox, random arrangement of buttons constitutes another unproductive load of discretionary information, in the form of novelty. The numbered buttons are novel because they are not arranged in the customary left-to-right and top-to-bottom order. The arrangement is also novel because the message buttons aren't grouped separately or

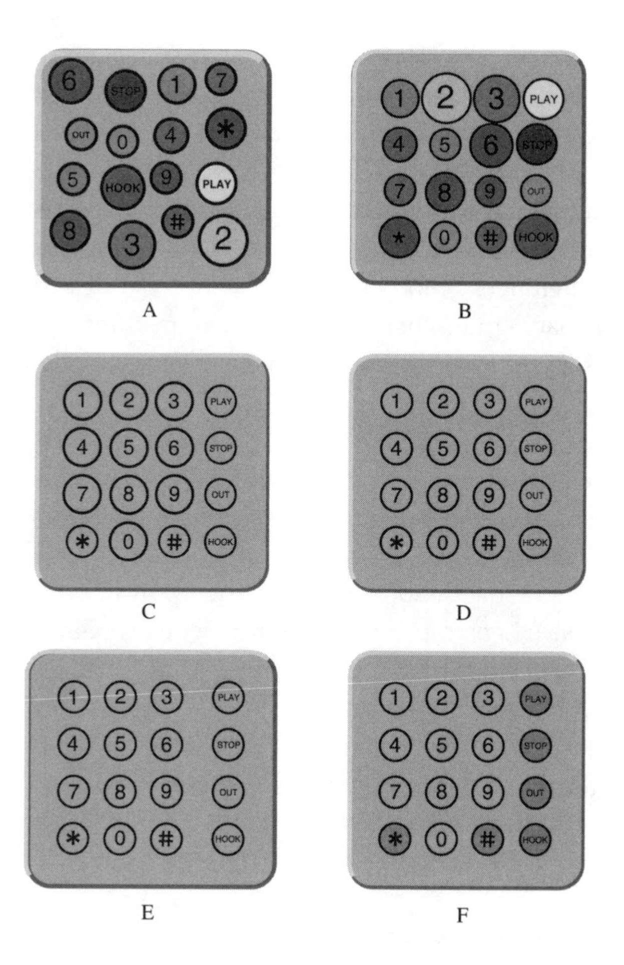

logically. Design A would require more mental effort and time to use than a normal layout, and the user would be more inclined to commit errors.

Design B is better because the left-to-right, top-to-bottom order of the numbered buttons conforms to the most expected arrangement, what ergonomists call a *cultural stereotype.* It would be easier to use, but the varying button sizes and colors imply nonexistent differences of meaning or importance: The larger size of buttons 2, 3, and 6, for example, suggest that they are more important than other numbered buttons; different colors suggest different functions.

Most of the contrast that amounts to discretionary information has been removed in Design C. The slightly larger size of the numbered buttons serves as collateral information that helps the user distinguish them from other buttons.

Identical size, color, and spacing of the buttons in Design D remove the last vestiges of discretionary information—to the extent that it might have too little information. Functional differences among the buttons are less obvious. In Design E, a difference in spacing (another instance of contrast) rectifies this problem. Different colors for different functions achieve the same end in Design F.

Such budgeting is apparent in many other products. At the business end of the iMac, where the screen displays a lot of essential and collateral information, the designer exercised very little discretion. As a result, it looks simple and rather conventional. If the designer had made the screen round, oval, or triangular, the computer would have been much more interesting. However, an unusual and inappropriate shape would have interfered seriously with the screen's ability to display a Web page and the user's ability to process essential and collateral information. The bulk of discretionary information shows up around the back and on the sides, where very little essential or collateral information exists. Consequently, the iMac's daimon is most evident in views from the rear.

Cars, especially their exterior surfaces, lie at the other extreme. They incorporate very few sources of essential or collateral information: taillights, brake lights, turn signals, and side-marker lights. Headlights illuminate sources of essential information in the environment for drivers. However, they do not provide essential information for viewers of the car, only collateral and discretionary information. In a collateral sense, they provide oncoming drivers with a helpful sense of the car's width and how far away it is; headlight shapes play an important role in the car's empathic expression—a matter of discretionary information that is more important aesthetically than in any practical or ergonomic sense. The designer has the discretion to make them seem "happy," as on the face of a New Beetle, or "scowling," as on the menacing face of an aggressive sports car or SUV. The preponderance of discretionary information in the exterior forms of cars probably explains why they are so much more interesting to design and why car designers prefer exterior design to interior design, where they have less discretion.

NOVELTY

Technically, novelty is a matter of contrast too, but novelty depends on *time-bound collation* of old and new, past and present. Because the only way we have of seeing the past is through memory, novelty depends on a mental model others cannot see or measure directly. Consequently, we have to call novelty *subjective information.*

Reduction of uncertainty with respect to novelty also depends on there being some uncertainty to reduce in the first place. This means that something unexpected, unknown, or unfamiliar has to occur. If someone says, "I have some information for you," you expect to hear something *novel* that you have not already heard—something *new* that resolves some uncertainty—such as whether you got the job you applied for yesterday. There are only two possibilities, of course: You either got the job or you didn't. Again, in terms of information theory, this is a situation with just two possibilities and the least possible amount of information, one bit.

If someone has taken the surprise out of the situation by leaking the news, the message would bear no information—and no ability to reduce your uncertainty in the matter—because you no longer have any uncertainty to dispel. The official acceptance letter can no longer surprise you because you no longer have any uncertainty to reduce. A message has to be *new* to you to be *news.* Yesterday's *news*paper is ho-hum "*old*paper" if you have read it already.

Remembering the Future

Perceiving novelty depends on having expectations and, therefore, a sense of the future. Our senses provide the closest thing we have to a sense of the here and now—but not actually. We cannot feel and know precisely the razor's edge of the *present.* As we can see trying to watch the hundredths of a second go by on a digital stopwatch, the nervous system is too slow. What seems to be happening right now, at this moment, is already history. Our awareness from moment to moment actually consists of mental fabrications from stuff dredged up from memory. It cannot be otherwise if you think about it: Sensory nerves take time to fire, signals consume more time traveling to the brain, and the brain takes yet more time to process, interpret, and register the event in consciousness. Although the whole process may take only milliseconds, awareness of the event occurs that many milliseconds after the actual event—after a new present has taken its place and gone into history itself.

It would be futile to even try to respond to the present. By the time we are aware of something happening to us, it has already passed. This is true even of the quickest reaction we can pull off, the knee-jerk spinal reflex that bypasses the plodding brain entirely; by the time it executes, the doctor's mallet has already struck. Consequently, the nervous system's best strategy is to plan ahead, to predict the future. Having already tripped on the curb and already heading facedown toward the cement, nothing is more important than what's about to happen. *The future always counts most.* Our nervous systems have been prepared genetically to "know" the future almost as well as the past—by mentally projecting the past forward as a "most probable" prediction of the future. I can predict, with certainty, that I am going to hit the ground. And although my brain is relatively slow, it is quick-witted enough to form a plan for minimizing the effects. I will decide during the second or two of the fall that I will put my hands out in front to break the fall—but not stiffly enough to break them—and roll to the side to minimize the risk to my face and head. I even recall, from playing sandlot football as a kid, that I should try to roll as I hit (actually, none of this was to any avail when I tripped and fell last year; I broke my shoulder anyway).

We predict the immediate future more accurately than the distant future, of course. You might not have a clue about next year, but you probably are fairly certain about tomorrow—or at least you have already constructed much of your prediction about tomorrow. Much of a successful maturing process involves honing your prediction skills and projecting your predictions farther into the future—with increasing accuracy. The more skilled you get, the longer you get to mature.

Memory's most important purpose—and why it almost surely evolved in the first place—pertains to the future and not, as we so blithely assume, the past. Nostalgic reflections aside, the future is a much more serious and fruitful thing to concentrate on than the past. Indeed, dwelling on the past to the exclusion of the future is one definition of neurosis—nothing can be done about what is already done. Anticipating the future accurately and reacting to it effectively—before it slips through the present to become part of the past—is the only effective way to cope; it is better not to trip on the curb in the first place.

Our predictions are extrapolations of that mental model of the world referred to earlier that Kenneth Boulding calls the *Image.* Abstractions, as well as direct experience, contribute to the Image. Although you may not have visited China, you could nevertheless have a good "China" component in your Image by virtue of things you have read about it (and thus imagined) and seen on TV or in movies. Regardless of how closely the Image matches *reality,* it is the springboard from which you jump to your conclusions and cast your bets about the future—whether of the next millisecond, the next minute, or the next year. It forms the basis of your plans but also your perceptions of the present. As the reality unfolds, it automatically brings to mind aspects of the Image that resemble it most closely. Maybe *dèjá vu* experiences are instances where an experience matches very closely that part of the Image it recalls. However, nothing matches the Image perfectly. Most recalled memories are incomplete abstractions, comprising only the high points of experience, things that seemed most interesting, most pleasant or most painful, and most important when first experienced.

Memories comprising the Image also seem unconstrained by the normal flow of time. The nostalgic feelings that flood the mind of a previous VW Beetle owner on seeing a New Beetle do not emerge as video sequences but all at once, all of a piece, to then imbue the perception of the new one. The first time you go to China, your expectations and initial perceptions of it—as well as your reactions to the actual experiences you encounter there—can only be based on your as yet artificially derived mental model of it.

To the extent that a place differs from your image and expectations the first time you visit, you will experience a shock to the system, the shock of *nov-*

elty. This is all that novelty is—a mismatch between the real thing and your expectations as determined by the Image. The extent of the shock will depend on the magnitude of the difference. As you return to the hometown you haven't seen for decades, you will be shocked by how much it now differs from your Image. If you were small when you left, the place will seem smaller now because your image will be based on a viewpoint, closer to the ground, that made everything seem larger when you were a child.

Stereotypes

By simply mentioning *watch* here, I have prompted your mind to shine its spotlight of attention on a small part of your Image—a generalized model of all the watches you have ever seen. I will call this a *stereotype.* Because the term has such abhorrent connotations of racial and other forms of social prejudice, I have been tempted to adopt another term. Other theorists use *schema* or *prototype,* but I find them wanting. A *prototype,* for example, refers to the "first of type," not "general type." So I keep returning to *stereotype,* which my dictionary defines as "a conventional, formulaic, and oversimplified conception, opinion, or image." It represents your best prediction, for the moment, of what the next watch you see will look like, an expectation. The Image consists of thousands, if not millions, of other stereotypes: of China, of hometowns, of cars, of teacups, of engineers, of houses, of computers, of cell phones and regular phones—anything and everything important enough to cause you to establish a mental category, with a name attached, so that you can remember it. Your watch stereotype has been shaped to some extent by every watch you have ever noticed but chiefly by those you have seen most often and most recently. Thus there is a good chance that your watch stereotype resembles most the watch you are wearing right now.

It is the mental image you would likely refer to if someone asked you to draw a typical watch from memory—but not necessarily. If, for instance, you wear a digital watch, you might regard it as atypical; you might choose to draw a more typical analog watch, more like most other people wear, with a round face, two hands of unequal length, and twelve numbers. However, details of color, size, finish, and materials (whether it has a leather strap or a metallic one, for instance) might differ depending on your personal experiences and preferences.

We prejudge the next watch we see on the basis of our watch stereotype—and tend to condemn it out of hand if it differs from the stereotype, just because it is "abnormal"—unless we find countervailing evidence that makes us change our mind.

The atypical, counterclockwise watch shown here on the right seems novel to the extent that it differs from your watch stereotype, which probably resembles the one on the left more closely. The novel watch requires more time and effort to tell time with because the novelty brings with it an additional information-processing load over and above that of the normal configuration. Someone using it also would be more prone to mistakenly interpret the time. The additional information-processing effort required to tell time with the watch on the right is due entirely to additional information in the form of novelty. Both images contain the same amount of information in the form of contrast; both images require the same number of bytes for storage in my computer's memory.

A normal watch compared with a novel, counterclockwise watch

Information and Aesthetic Potential

Like all things biological and behavioral, the Image and all its associated stereotypes exist only because they have proved beneficial to prolongation of life and survival of the species. The more completely and accurately they model the real world, the more likely we will make accurate predictions in anticipation of threats and benefits. A good match leads to a good prognosis: "If I survived something like this before, I'll probably survive this time." Not much to get excited about.

The prognosis associated with the poor match of a novel situation—implying that you might not know how to deal with the situation—*is* reason for concern and even preparation for the worst. The resulting agitation of emotions, thoughts, and urges, called *arousal,* is the progenitor of the so-called fight or flight response characterized by fear and anger—so crucial to survival in truly threatening circumstances—but also the aesthetic response to art. The heightened state of attention and awareness accompanying arousal—called the *orientation reaction*—marshals all the available senses in an effort to decide whether to fight, flee, consume, or ignore the arousing stimulus.

MEASURING NOVELTY

We can measure a design's novelty and, by implication, its aesthetic potential by comparing it with its stereotype. We can tap the stereotype in at least two ways, visually and semantically. We can construct a semantic profile of a stereotype in the same way we would construct a profile of its ideal (see the semantic profile of the ideal watch compared with Watches A and B on p. 64). Instead of imagining "the ideal watch" as they filled in the semantic differential form, subjects would imagine an equally abstract concept, "the typical watch" (the stereotype). Profiles of ideals and stereotypes usually differ: Stereotypes signify what subjects expect a product to look like; ideals signify what they hope it will look like. The distance between the profiles of a real product and its imaginary stereotype amounts to a measurement of the product's novelty. A product that corresponds closely to its stereotype seems quite ordinary. It seems novel to the extent that it differs from its stereotype.

Visual Stereotypes

I suggested earlier that someone's drawing of a "typical" watch, done strictly from memory, probably would resemble their watch stereotype as much as any particular watch. Presumably, drawings from a group of people, averaged together, would yield an even more general stereotype that represented the group as a whole. I have tested this hypothesis successfully with respect to such complex products as cars. However, for an example, I will show the results from a study concerning a much simpler product, an ordinary teacup.

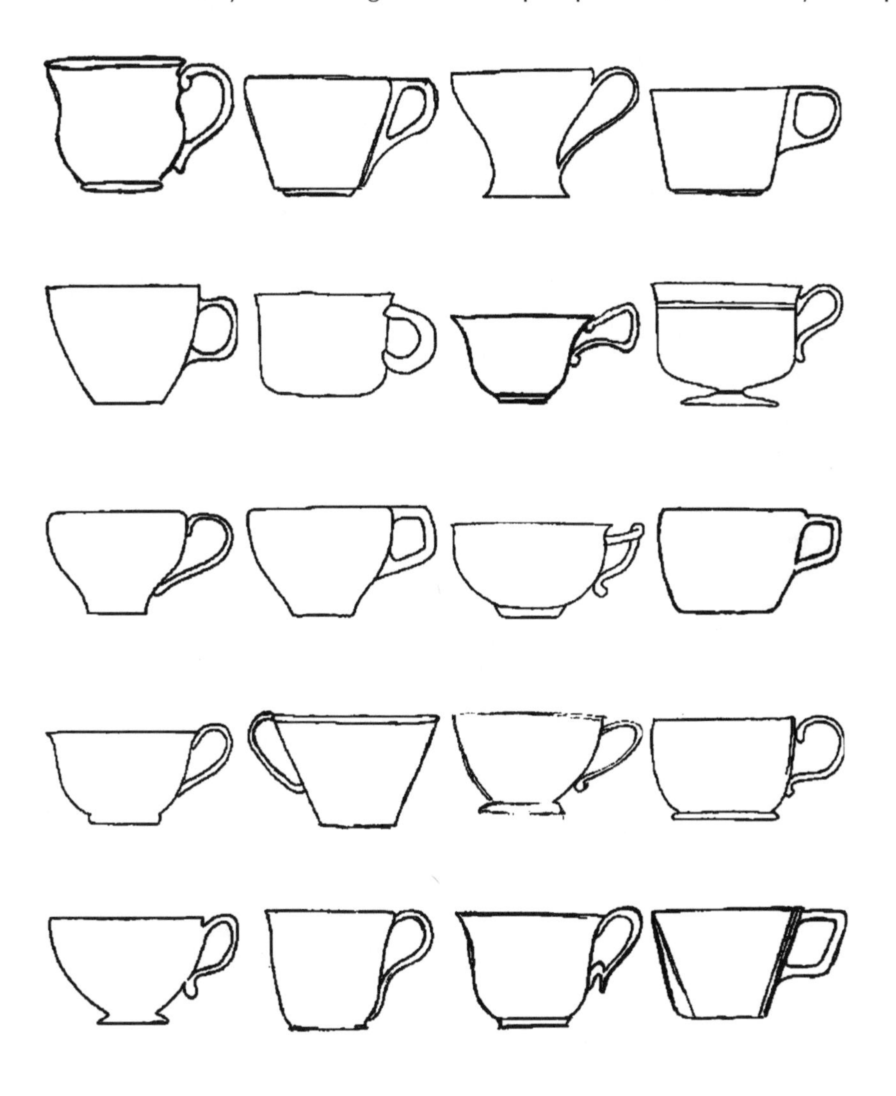

I asked 20 students in a beginning product design course to each "draw a full-sized side view of a typical teacup." I stressed that this was not a design exercise: They were not to design a cup but only to depict a typical cup, the first thing that came to mind. The drawing in each case probably reflected the cup that the student used most frequently, probably the one he or she drank coffee or tea from that morning.

Note, incidentally, how your eye soon goes to the unique cup with its handle on the left. The person who drew it probably was left-handed, so he or she was used to seeing handles that pointed to the left.

Each profile has a relatively unique form, as expected, yet they collectively reveal commonalties that hint at a more general stereotype: Most are widest at the top, most handles are near the top, and height-to-width proportions are remarkably consistent. Without further analysis, these consistencies suggested at least three ways a designer could achieve novelty and aesthetic potential in a new cup design: Make the bottom larger than the top, place the handle nearer the bottom, or make the cup taller or wider than usual.

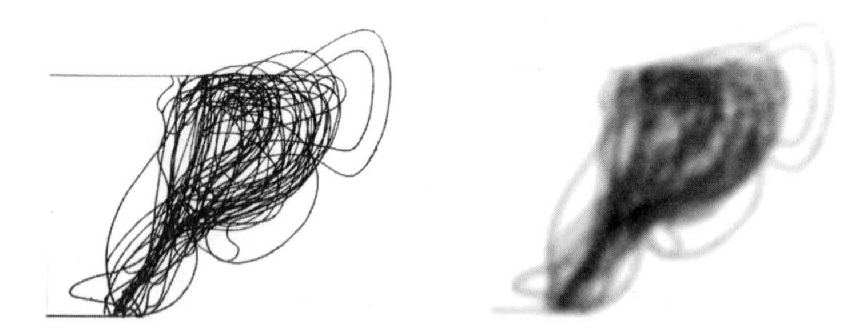

You can begin to get a sense of the stereotype from the compilation of all 20 cups shown here (the left-pointing handle was moved to the right). The most novel features also stand out clearly from the crowd.

Throwing the image out of focus causes a more definite, if ghostlike, image of the stereotype to emerge. Profiles near the norm reinforce each other where the image is darkest. The less normal ones receive little or no reinforcement and tend to fade.

Corning Corelle cup compared with computer-generated stereotypical cup

The classic Corning Corelle cup shown here closely resembles Eric Chen's computer-generated stereotype—the statistical mean of all 20 drawings—except for its unusual "beaver tail" handle. The handle departs from the stereotype enough to stand out as a novel and eye-catching feature. It evokes the orientation reaction and becomes the center of attention. The handle dominates not only the viewer's gaze, but her thoughts, feelings, and urges—especially the urge to reach out and test the handle's authenticity. The viewer will find it wanting—simply on grounds of its abnormality—unless and until it makes sense.

RELATIVE USES OF CONTRAST AND NOVELTY

Both novelty and contrast constitute aesthetic engines, directly responsible for sensations, feelings, and thoughts—ranging from subliminal to profound. Because the nervous system responds to information—and only information—novelty and contrast are the *only* means by which a product can grab a viewer's attention, get the emotional juices flowing, start the cognitive wheels turning, and urge the viewer to action.

At the beginning of this chapter I noted that, although both contrast and novelty involve comparisons of stimuli, contrast is essentially objective in nature, while novelty is subjective. This means that you can measure contrast directly by simply measuring and comparing attributes of the object itself. Measurements of novelty are more involved and less certain because they require tapping the viewer's mind for a relevant stereotype.

Another difference is important. While contrast remains constant over time, novelty never lasts. The level of contrast perceived in a watch today will seem the same tomorrow and produce the same effects. The novel watch, however, having been seen, is never again as novel as it once was. Even if you never saw a backwards watch again, recollection of it would never seem as novel. If you began wearing one and using it on a regular basis, it eventually would dominate your watch stereotype so thoroughly that it would seem quite normal—to you, if to no one else. Having reshaped some details of your brain's structure, it would have *informed* it with new *knowledge.* Having *learned* something in the process, you could now tell time with it as readily and accurately as with your old watch. In fact, if you were to return to your old watch, you would find it briefly more difficult to use than the new one. In effect, the stereotype would have swung so far toward the counterclockwise configuration that any clockwise watch would seem novel—and thus more difficult to use. Stereotypes are resilient, however, and soon enough the old watch would seem normal again.

In describing contrast, I noted a counterbalancing effect: As we *increase* the uncertainty of a ruler by increasing the number of its increments, we *decrease* the mental uncertainty of the ruler's user—until, at some perceptual limit the user's mental uncertainty begins to increase. No such reversal occurs with novelty: Increasing a product's novelty *always* increases mental uncertainty.

This means that

- Contrast is more important as a source of essential or collateral information, where immediate reduction of uncertainty is ergonomically important:
- Novelty is especially important as a source of discretionary information, where elevated states of arousal are aesthetically crucial; novelty is the aesthetic engine of choice in product design.

The four-year sequence of Pontiac GTOs shown on the next page illustrates how the American automobile industry traditionally extended a design's useful life by invigorating it with annual doses of information called *facelifts.* Each facelift made the design more interesting and exciting through additional contrast and/or novelty.

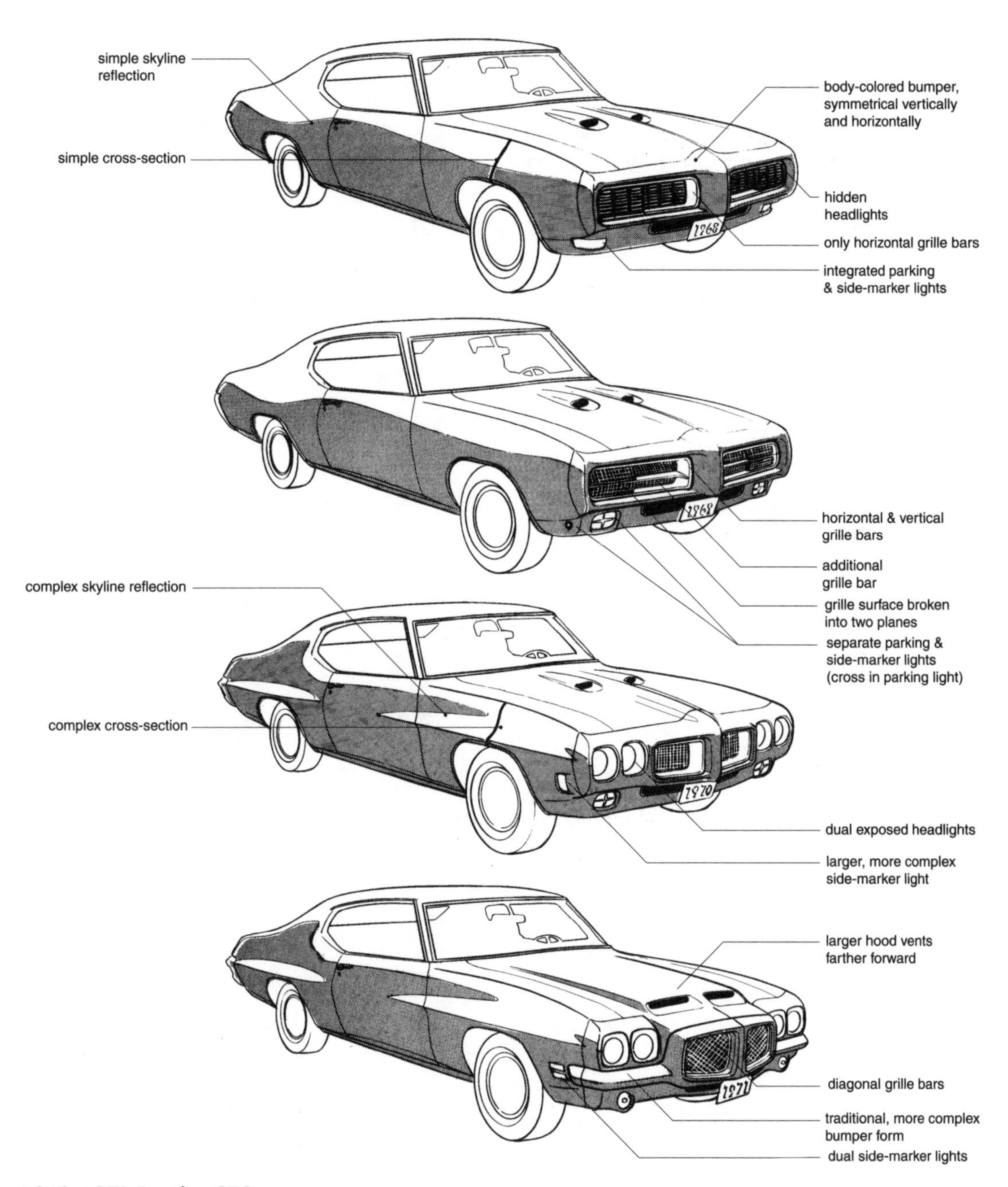

1968–1971 Pontiac GTOs

The novelty of the initial 1968 model lay, ironically, in its sparse contrast; it was unusually simple and clean when compared with its predecessors and competitors. This was evident in its "horizon reflection," the reflection of the bottom edge of the sky in the car's mirrorlike surface; it meandered the length of the car with few sudden or extreme changes of direction. (This feature is as important a design feature as the car's profile or any other perceivable feature.) Simple cross sections (which describe the relatively simple geometry of this car's side) contribute to the reflection's simplicity. The body-colored plastic nose, an industry first, also was novel; so was its unusually simple shape and symmetry (not only side to side, but top to bottom). Hidden headlights and a simple grille texture composed of just horizontal grille elements (no vertical elements) maintained the simplicity. So did integrated parking and side marker lights.

By the time the 1969 model was introduced, much of the novelty of the original model had worn off because nothing can remain new forever. Designers introduced additional contrast by

- Adding vertical elements to the grille texture
- Adding a prominent horizontal bar across the middle of each half of the grille
- Creasing the grille texture and headlight covers to give them a slight V-shaped cross section
- Splitting the parking and side marker lights and adding an ornamental cross to the parking light and making it protrude from the surface

Changes for the 1970 model, called a *major facelift,* were more extensive and costly because, by then, the 2-year-old body had grown quite familiar. The side became more complex by virtue of more complex cross sections; the one through the front fender now had a change in direction of curvature (an inflection) that increased its information. Each change in direction of curvature introduces the likelihood of *two* additional horizon reflections and a disproportionate increase in complexity. This is evident in the front edge of the door, which now crosses three horizon reflections instead of just one. Where the horizon reflection does not break up into three, toward the front of the fender, it gains contrast by undulating sharply back and forth. The pattern of reflections becomes even more intriguing when the car moves by the viewer, or the viewer moves, because the reflections also move. The more complex the surface, the more animated the reflections become; again, this adds to their aesthetic potential. The designers also increased contrast by:

- Uncovering the headlights and making them separate, prominently recessed elements
- Changing the side marker light from a simple round shape to a less regular shape

For 1971, the final year of production, the designers increased contrast by

- Doubling the number of side marker lights (notice, however, that they have gone back to a simpler round shape)
- Making the grille jut farther forward
- Reverting to a more traditional and complex bumper form, rather than the integrated nose
- Orienting the grille texture at a 45-degree angle (technically, this might be considered novel rather than a case of contrast)
- Enlarging the hood scoops and moving them forward (this increased arousal potential by virtue of exaggeration)

Each facelift resulted in a design that seems more active than the last. When comparing the last with the first, it is noticeably more active. Considering the emotional-rational scale of the semantic differential, the design moved progressively away from the rational pole toward the emotional—just what we should expect as information increased.

Chrysler designers did not want to depart too far from the look of their market-leading minivans when they redesigned them for 2001. So the new one looked much like a facelift of its all-but-identical predecessor. The designers used contrast to "freshen" the minivan's appearance. The contrast shows up in the crisper details in the hood, grille, and sides and in the exaggerated shape of the taillights.

2000 and 2001 Chrysler minivans

CLASSIC DESIGN

The tactical advantages of novelty in the marketplace are obvious: It attracts and holds the attention of consumers while evoking aesthetic responses at the expense of competing products. Longer-term, strategic advantages—such as increasing brand equity—are more likely to be gained by using special kinds of novelty usually found in *classic* products. Essentially, a classic product departs from norms so much that it no longer fits neatly into any existing class. It consequently requires definition of a new class if it is to be categorized at all—thus the term *classic.* Classics are often marked by the following:

- What William Abernathy calls *epochal innovation* has more enduring value than mere novelty. A product with epochal innovation fills a practical need or solves a problem not yet met or does so in a markedly better way than previous products. An epochal innovation establishes a new *practical* standard for a class of products, which compels competitors to emulate it in order to stay competitive.
- The *seminal form* of a product similarly establishes a new *visual* standard for a class of products and seeds a new trend that competitors are obliged to follow. Seminal form is original form by definition. It also expresses *authenticity.* Whereas the authenticity of epochal innovation arises from functional improvement, the authenticity of seminal form arises from what *seems* to be; it is strictly aesthetic in nature. To seem authentic, a product's appearance must empathically suggest that it "conforms to fact," as when its form seems consistent with its function. An authentic product does not lie. It does not seem contrived or phony but honest and worthy of confidence—just as an authentic person does.

The significance and competitive advantages of seminal form and epochal innovation cannot be ignored by competitors; either one forces competitors to emulate it in order to stay in the game. Imitation in this case is not just the sincerest form of flattery. We always remember the first instance of anything best, whether it is the first date, the first child, or the first product of a kind. We remember less well, or not at all, the instances that follow. By adding to the ranks of the newly established class, each imitator confirms and reinforces the leadership, uniqueness, and authenticity of the original—and its brand.

The lasting appeal of classic design translates into greater brand equity, increased corporate prestige, more resilient customer loyalty, longer product life cycles—and greater profits.

Movado Museum Watch

The classic status of the Horwitt watch (shown here in an early "Museum Watch" interpretation by Movado) was not earned because the watch was an example of epochal innovation. While Horwitt endeavored to show that one could tell time with very few sources of explicit essential information, the watch's elegant design did nothing to improve the user's ability to tell time. The design earned its status, instead, on aesthetic grounds, by virtue of the seminal innovation of a numberless face. The face's black color was novel as well, and its sharp contrast with the metallic case helped to make it memorable. As proof of the Horwitt's seminal nature, it has become one of the most memorable and emulated timepieces of all time.

In contrast to the Horwitt watch, the first production hybrid car, the Toyota Prius, qualifies as an epochal innovation. With a driveline that smoothly

2001 Toyota Prius

blends power from an electric motor and a gasoline engine, the car's breakthrough engineering achievement establishes its potential for classic status. Its rather conventional exterior will not earn a reputation as a seminal design, however.

1995 Mercury Sable

The 1995 Mercury Sable provides an example of design that is both epochal and seminal. Originally introduced in 1986, along with the very similar Ford Taurus, its improved aerodynamic form qualified as an epochal innovation by improving fuel efficiency at no additional cost at a time when fuel prices in the United States were rising rapidly. It achieved status as a seminal design by establishing the "aero" look that still sets the standard for automotive design. Despite the fact that this 15-year-old design looks quite ordinary now—in part because it has been emulated so widely—its "jelly bean" form was novel to a daring degree in 1986, when the model it replaced and its competitors had boxy shapes. It made so much sense, functionally and aesthetically, that it sent every other manufacturer back to the drawing boards. The new look brought Ford Motor Company back from the brink of bankruptcy by generating some 400,000 sales the first year. By resetting the norm for automotive design overnight, it established Ford as the design leader among American carmakers for several years.

Classic products ultimately provide the most enduring pleasure and profits because they never go "out of style." The appeal of such products remains so immune to shifting fads, fashions, and tastes that they often can remain in production indefinitely, providing continual pleasure and profits while boosting brand equity. For industrial designers, classic potential is the Holy Grail, the objective for which their instincts and training have prepared them. I do not know of an industrial designer who doesn't yearn to have his or her work chosen for an art museum's permanent collection. And manufacturers should seek it, too, for it can add long-term value to a product and its brand at little or no additional cost.

I have now described the two qualities—novelty and contrast—that are both necessary and sufficient for initiating all aesthetic reactions. They excite our nervous systems, pique our attention, stir our emotions, and move us to act. No combination of novelty and contrast can guarantee that a product will be beautiful, however; ugly products also embody one or both of these forms of information. The issue of psychological valence—whether we like or dislike something—depends on how much concinnity the stimulus also happens to have. We will turn our attention to matters of liking and disliking in the chapters that follow.

CREATIVE CAD

Obviously, a computer can be used to modulate a design's contrast. The one on which I am writing these words can do so quite simply by changing the brightness of the display's background in response to my command or automatically in response to prescribed circumstances. Using its CAD software, it also can bend a straight line or increase the curvature of an arc to increase shape contrast.

Novelty is a knottier issue because it is subjective in nature. Since you and I have different expectations, what is novel and exciting to me may be familiar and ho-hum to you. Designers have a hard time conjuring up truly original designs. For one thing, its hard for us to shake the influence of what we already know, to go counter to it. Designers have a strong tendency to base anything they design on something they have already designed or seen.

A computer has a head start in this regard because, fundamentally, it is too "dumb" to know anything about its history as long as its programmers keep it in the dark. Left to its own digital devices, it has nothing "ordinary" from which to extrapolate. It has a second advantage in its tireless ability to be prolific. It can generate an endless series of permutations of a design. This talent, coupled with the parametric design technology described earlier, gives the modern CAD system the potential of an automated "creation machine" that could leverage a designer's creativity capacity enormously.

Let's consider the possibilities with a trivial but illustrative example, an ordinary glass tumbler. Imagine that this design project begins like a lot of them do, with a sketch on a napkin during lunch. Despite the approximate nature of the sketch, the designer and her colleague across the table understand that the hypothetical product would be essentially cylindrical, that it would have a certain height and diameter, and that it would hold a certain amount of water. But they do not have to consider specific values in connection with any of these parameters for the time being. They can concentrate on what they are primarily interested in, aesthetically relevant parameters such as shape and proportion.

The volume parameter for an ordinary, cylindrical glass tumbler, for example, depends on the product of three other parameters for its value: height, diameter (or radius), and the universal constant pi (π). Assuming that the designer wants any design to have the same volume, changing either height or diameter causes the other to assume a new value, too.

With a CAD system, the designer also begins by "sketching" a generic model of a tumbler in the computer's memory and specifying values for its "volume" (say enough to hold about 8 ounces) and "wall thickness" parameters, which will remain fixed. The universal constant pi (π), needed for calculating the volume of a cylinder, would constitute a third fixed parameter. However, pi is embedded in the formula for calculating a cylinder's volume, which the designer never sees. The formula works behind the scenes to ensure that when the designer specifies a particular height, the computer automatically chooses the unique diameter necessary to maintain the desired volume.

In the hypothetical case I am posing here, the designer does not specify height directly but establishes a range of values from the shortest practical tumbler to the tallest. These limits become two more parameters. Finally, she specifies the number of heights the computer should try in between the limits.

Following an algorithmic sequence of steps already programmed, the computer incrementally steps through 10 different heights in this example and calculates the required diameter for each by reconciling it with the required volume. It displays photographlike renderings of all 10 designs—all within seconds.

Even in this trivial example, the computer leverages the designer's capabilities immensely. It would have taken the designer much longer to produce a rendering of just one design using an airbrush, pastels, or a pencil. The time allowed for the project might not allow the designer to consider more than two or three different designs. Not only does the computer free the designer from having to calculate the diameter associated with each height, in order to determine the aesthetically important proportions of each design, the designer does not have to know the formula. Actually, the computer could have created a hundred designs, or a thousand, given a little more time.

If the designer had used traditional visualization media, then her client unavoidably would have had to compare apples and oranges in selecting the best. Inconsistencies in the rendering technique from one design to the next would make them, in the strictest sense, incomparable. Nuances of rendering might be enough to outweigh nuances of design. The client might end up selecting one design not because it had the best proportions but because it seemed slightly more realistic than the others. The computer, on the other hand, is virtually incapable of inconsistency by its very nature. Furthermore, the computer can show any or all of its designs from any view, in true perspective, while consistently maintaining realistic lighting and reflection effects—just as real tumblers would.

The computer improves the design process by effecting a more appropriate and satisfying division of labor: Even if the designer could have created 10 designs and 10 renderings in the allotted time, would she have wanted to do so? She'd be bored out of her gourd. The computer can allocate to the designer what she is best at and loves to do (such as creative inspiration, intuition, and judgment) and take for itself tedious, humdrum stuff it does best and seems to love (repetitious, methodical number crunching).

I kept the preceding example simple for purposes of explanation. However, it merely scratches the surface of parametric CAD's ultimate creative potential. The designer could have defined virtually any number of additional, aesthetically relevant parameters: color, transparency, shapes other than a cylinder, etc.

A computer created these nine tumbler designs by varying just one more parameter, pertaining to the "tangent vector" values at top and bottom, that determined how straight or S-shaped their profiles would be (if this parameter were set to zero, the resulting form would be a cylinder). The computer's labor-saving leverage becomes more obvious in this example. Calculation of

volume involves a more complex integral function, the kind the designer would have to have learned in a calculus class instead of junior high math. Creative potential increased, too. Assuming that the computer again stepped through 10 different heights, stepping through 10 different tangent-vector values would have increased the number of designs tenfold to 100.

Creative potential increases with each additional parameter. Assuming 10 steps per parameter, the number of possible designs increases tenfold with each additional parameter. Just count the zeros: A third variable parameter (different diameters for top and bottom, perhaps) yields 1000 designs; a fourth (different tangent-vectors for top and bottom, maybe) yields 10,000; define just six parameters and you have the potential for creating a million different designs!

As "creative" as the system would be, in the technical sense that it generated a lot of designs, it still might not deliver much novelty—the quality we specifically need. Each additional parameter increases the chance of something unexpected. But less than they might be as long as the designer continues to provide all the up-front instructions—deciding what to change and how much to change it.

Computers have inherent potential for generating new and unexpected results simply because they can be programmed to interject random values into the design process. For example, the computer might suddenly introduce a randomly determined alternative to the circular shape (such as a square) for either the top or bottom of the tumbler (which also might be determined randomly). The 10 different shapes developed by interpolating between the square and the circle would increase the number of designs another 10 times—as well as the chances that some would be unusual enough to be surprising.

Other inputs might alter results in unexpected ways. Just *how* would be limited only by the imagination of the people programming the computer. The computer could create a geometric analog of a note or chord from a musical instrument and use it to determine the overall shape of a set of tumblers. An F-sharp tumbler might turn out to be more novel than a B-flat tumbler. A tumbler informed by a Bach fugue probably would be more complex than one informed by a Beethoven symphony.

However, the computer's creativity and potential for novelty solve only half the problem at best. Indeed, the computer's prolific tendencies compound the other half of the problem—finding the one design that most *pleasantly* surprises viewers. While most of the computer's prodigious output will amount to flotsam, its prodigious nature represents its greatest asset. During its random walk through the vastness of the space containing all possible tumbler designs, it is more likely to stumble on at least one novel design than consumers would like—like the proverbial roomful of monkeys pounding away on word processors. Given enough monkeys, enough word processors, and enough time, one of the monkeys will write a sonnet to rival Shakespeare.

We need to get realistic, however. You couldn't display a million designs on the screen at the same time. Printing them on a million sheets of paper would be prohibitively expensive and time consuming. If you did, who would go through the dizzying job of digging through all that chaff to cull the designs worth serious consideration? Even if someone were willing, it isn't humanly feasible to sort through a million designs to find the one best design. Nor is it practical to sort through a thousand or, for that matter, even a hundred. Culling the winners would be a far more daunting task than creating the candidates in the first place.

What we need, of course, is a computer with "good taste." We wouldn't expect it to show us a picture of every concept it came up with. This would waste a tremendous amount of the computer's resources. Instead, it would keep all the designs to itself, stored on disk as compact "genetic" codes of ones and zeros, waiting to be expressed visually if need be. Agents armed with aesthetic rules would do the actual winnowing, at the computer's blinding pace. Working them in concert with agents responsible for practical and ergonomic issues, they would narrow the field down to a humanly manageable 10 or 20 best designs, with regard to all the constraints and objectives factored in—or even 1 design, if asked to do so. I will address the prospects for the rational aesthetic rules the agents would need when we consider the subject of concinnity in greater detail.

States of Mind and Body

7

CUSTOMERS, USERS, AND the public react to products through the mental and physical processes by which they think and feel. How they become aware of a product, how they generally feel about it, and how they physically respond to it largely define the product's empathic character. To the limited extent that it can be distinguished from feeling, thinking largely aims at determining the authenticity of a product. The aesthetically important processes—awareness, affect, arousal, conation, orientation, and cognition—interact within a unified mind and body.

Even the most mundane of products—an ordinary thumbtack—embodies the extremes of information found in the most complex product. By holding a thumbtack between your fingers and carefully squeezing it tighter and tighter,

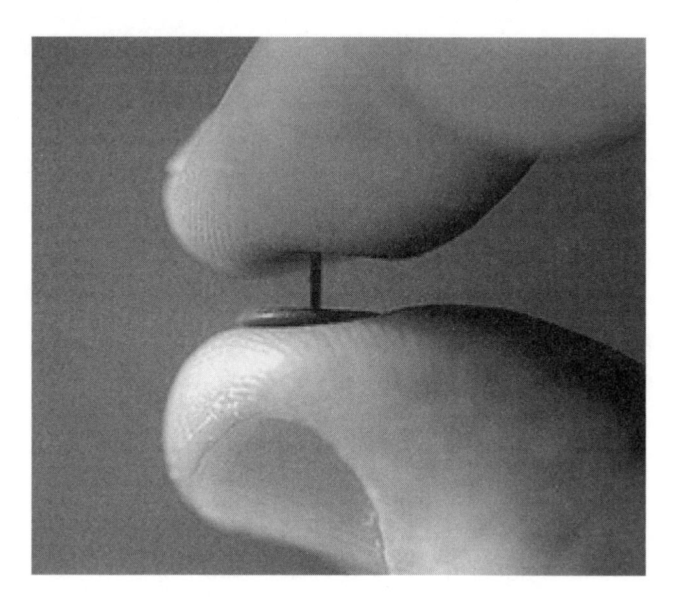

you can experience the entire gamut of what we know as *aesthetic experience*—sensations, directed and concentrated attention, emotions (or feelings), urges to act, thoughts, and appraisals ranging from indifference to good or bad. Sensations from the front lines of the nervous system, the senses, come to mind first. You sense the sharpness of its point and fix your gaze and thoughts on the point and nowhere else. You "mind" the tack's point first because, by virtue of the extreme variation of surface there, it embodies more information than anywhere else—especially the virtually flat and "formless" cap. You feel a growing anxiety, perhaps, and an urge to let go. The experience is interesting for a while—certainly more interesting than the cap—but eventually painful as you continue to squeeze. You might think to yourself, "Ow, that's sharp! Why am I doing this?" We experience all these states of mind—and body—collectively and empathically.

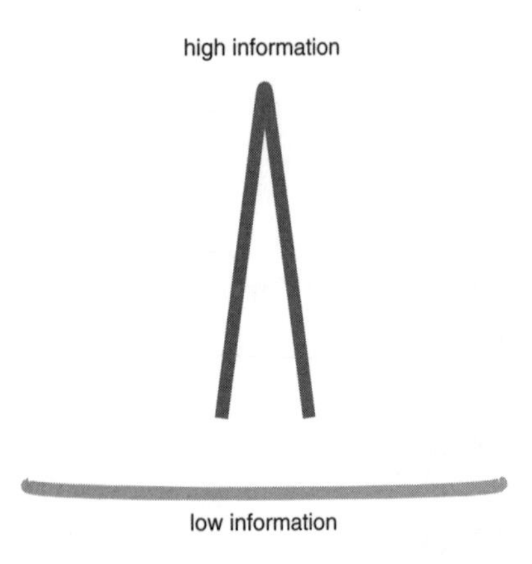

Sharpness, above all, characterizes a tack's daimon. However, all sensations involved in the experience of squeezing one, even vision, serve to flesh out the tack's persona. Thus a sharp break in a line (like the sharp break in the surface of a tack's point) draws more attention than a shallow curve that resembles the smoother surface of a tack's cap. Not only does the pointed figure hold your eye and attention more, it also increases general arousal, as if you had actually touched the point of a real tack with your finger.

All awareness and knowledge begin with the sensory nerves that cover the body's entire surface, inside and out. This system, composed chiefly of the skin and the lining of the gut, constitutes the interface between individual and environment. Some sensations acquired through the senses we are all familiar with acquaint us with sounds, smells, sights, tastes, cold and heat, and pressures of the external environment. Other sensations from the viscera—the soft organs of the abdomen and chest, especially the digestive system—add their mysterious input. Even constrictions and dilations of blood vessels add subtly to the overall experience. Together these amount to the emotional reaction to the experience. They constitute the most direct evidence of the tack's daimonic presence; the point of the tack not only looks sharp, it is sharp. It carries with it actual sensations but also potential pain.

AWARENESS AND CONSCIOUSNESS

We often use the words *awareness* and *consciousness* interchangeably when discussing our reactions to experience. As I understand them, however, awareness is a more general state of mind than consciousness and less intellectual. Awareness accompanies all behaviors, including the most instinctive, automatic, or intuitive. Consciousness seems to require thinking about what we are aware of.

Consciousness involves paying *attention* to something and the ability to recall it from short-term memory. This issue reminds me of a scene in an episode of "Perry Mason" I saw many years ago. Testing the reliability of a witness' ability to identify his client as a suspect, Mason turned his back to the witness and asked, "What color is my tie?" The witness could not answer because, although he had surely seen the tie, he had not "paid attention" to it; he had been *aware* of it without being *conscious* of it.

When I drive my familiar route to work, I am *aware* of the environment and the traffic, without necessarily being *conscious* of them. When a car routinely merges into my lane ahead of me, measurements of my brain's activity would reveal patterns consistent with perceiving the event and slowing my car to make room for the other or turning the wheel to change lanes—as well as glances in the rear-view mirror. However, as long as it does not emerge suddenly and unexpectedly (the very meaning of an *emerg*ency), I remain no more conscious of details, such as the car's license number, than the details of habitual tooth brushing. Having committed a detailed map of the route to memory, I switched on my autopilot as soon as I entered the freeway. To be conscious of the trip, I would have to take an unfamiliar route. To be conscious of brushing my teeth, I might hold the brush in my other hand.

To become more conscious or "mindful" of something entails an increase in cognitive activity. When we ask someone's permission with "Do you mind?" we are asking them if some new intrusion we want them to attend to bothers them by burdening their cognitive resources. It is an important courtesy. Greater devotion of cognitive capacity to something does not guarantee better performance. To begin with, to *attract* the mind to something new *distracts* it from whatever else it was occupied with. Try thinking about brushing your teeth or walking, and you will do neither as smoothly, effortlessly, or skillfully as when you do not think about them. Nor will you be able to think of anything else. In my own case, I can spend my consciousness on other, more pressing or interesting things as I brush my teeth or drive the route to work. Both are so deeply etched in memory that I can think about the lecture for my first class, a committee meeting afterward, or what to get my wife for her birthday. I don't have to "wake up" until I reach the parking garage, where I always need all my wits to find a space. I would have been so *unconscious* until then that I would be hard-pressed to describe any of the familiar things along the way—even though I had been *aware* of them.

An aesthetic experience arises from contact with the real world. It always involves attention and therefore consciousness; to be moved by something is to be conscious of it. Plato would have taken exception to this. He proclaimed that the material things you and I think of as real are not really real. In his view, reality would be comprised instead of fundamental mental abstractions or models he called *ideals* or *forms.* A real (i.e., *perfect*) circle, for example, could exist only as an intangible mental ideal. Peer through a magnifying glass at a circle drawn on paper with a pencil or a computer and you will see what Plato meant. It would not consist of "an infinite number of points all lying equidistant from the center," as the definition of a circle requires. Regardless of how carefully drawn, the material circle has defects the mental model does not have. Even if the circle were drawn with an extremely rigid and precise compass, the graininess of the paper would turn the penciled circle into only a ragged approximation of the mental ideal. The points might lie, on average, the prescribed distance from the center, but few would lie exactly at that distance. And the finite number of pixels (picture elements) in a computer image—necessarily composed on a rectilinear grid—ensure that the points of a computer-drawn circle cannot all lie equidistant from the center.

If aesthetics concerned material things, it followed that Plato's immaterial ideals had to be anesthetic in nature. Thus, to this day, we understand *platonic love* as adoration without sensation or affect, devoid of emotion, passion, or lust. Plato had so little respect or trust for poets and other artists who dealt in aesthetic matters—who trafficked in affect, affection, and affectation—that he banned them from his imagined ideal republic.

In an ironic twist, Plato's best-known student, Aristotle, turned his teacher's notion on its head to give us the concepts of the real and ideal most widely held today:

- Material objects we can see, hear, smell, taste, and feel directly through our senses—aesthetic things—are the real things;
- Immaterial, mental images, concepts, ideas, and ideals—which cannot directly stimulate the senses except through recalled memories—are only figments of imagination and decidedly unreal.

Thus a tack's point is more aesthetic than its cap because it produces an effect that is "sharper in the senses." The point informs the person squeezing it that it embodies more information than the cap. By the same token, it also has greater potential for *informing* (or *deforming*) something else. You can see this in the way it depresses the skin more than the cap does. This differential of information accounts for the tack's functional and ergonomic appropriateness; properly used, the point of a tack can be pushed easily into a bulletin board by virtue of its concentrated information—without hurting your thumb by virtue of the cap's diffused information. If you were foolish enough to try pushing the tack in cap first—expecting the low-information cap to penetrate the board before the high-information point penetrated your thumb—you would be sadly and painfully disappointed; you would end up with only an ergonomic problem for your trouble. Language reflects these facts metaphorically, such as when the teacher asks, "Do you get my point?" after trying to make an "impression" with a "penetrating" explanation. The students who get the point are "sharp as tacks"; those who do not are "dull" as a tack's cap.

An appreciation of the elegant coupling of functional and ergonomic objectives in such a simple and inexpensive product might lead you to enjoy its consequent intrinsic beauty, especially the next time you mount something to a bulletin board; its form makes sense. This is the essence of that old chestnut among designers that "form follows function." A product's beauty is doubly appreciated when its form follows not only functional but also ergonomic callings. However, what we like we also can dislike under different circumstances; if you accidentally stab your finger while probing the box for another tack, you will think less kindly of tacks, at least momentarily.

The sensations that initiate awareness are joined quickly by five other states of mind and body—*affect, conation, cognition, appraisal,* and *valence.* They blend inextricably with the sensations to become part of the consummate mental image of a product (or any other stimulus) that is stored in memory for future reference.

AFFECT AND AROUSAL

Affect—which is used interchangeably for *feelings, emotions,* and *moods*—traditionally has been the catchall for all those states of awareness we do not think of as thinking. It is the irrational side of awareness, if you will, that accounts for uncontrollable, illogical passions. It also has been associated with a pleasantness-unpleasantness or hedonic dimension. At any rate, affect comes to mind very quickly, right on the heels of sensations.

Prototypical fear is first and foremost among emotions. The nervous system evolved to be a master at the game of playing it safe, and fear is the safest of all emotions. A tack presents real cause for fearful concern, of course; it actually can hurt you. However, we are so conditioned to playing it safe that fear is instinctively the first—if fleeting—feeling to emerge. This is important aesthetically because fear is the most energetic and motivating affect. It lies at the heart of the fundamental urge for flight or fight. Other, less fearful feelings well up from associated memories of past experiences with tacks (perhaps even the first time you used one in kindergarten) or similar things (the Washington monument, perhaps). In most cases, they are so vague and blurred by intervening experiences that they cannot be distinguished. They just blend with all the rest, present and remembered, as they bind with the look and feel of this particular tack. Thoughts and images, which are cognitive in nature, all mix in. The imaginary stereotype, representing all thumbtacks in general, is one of the most important of these.

The heightened awareness of the tack entrains a raft of reactions throughout the body, under control of the sympathetic branch of the autonomic nervous system—some of which provide the only objectively measurable evidence of aesthetic experience. This complex mix of psychological and physiologic reactions, technically known as *arousal,* manifests itself through many objective and measurable components:

- Changes in blood pressure and heart rate
- Dilation of blood vessels of the extremities
- Changes of breathing rate and pattern
- Expansion of the pupils of the eyes
- Decreased electrical resistance of the skin (galvanic skin response, or GSR)
- Changes in electroencephalographic (EEG) waves (increased frequency, decreased amplitude)
- Physical and chemical changes within sense organs that increase their sensitivity (lowered sensory thresholds)
- General restlessness
- Elevated general muscle tension

An aroused person wired to a polygraph (more commonly known as a *lie detector*), which graphs several components of arousal, would cause the pens drawing the graph to swing more widely and wildly than usual. A blood sample would show increased amounts of epinephrine and other hormones secreted during arousal. The consequence of arousal is a temporary increase in physical and mental prowess; the aroused person is stronger and smarter—a better processor of information. Arousal is the signature of the classic fight or flight syndrome that prepares an organism for potential emergencies. Arousal is the body's acknowledgment that the situation at hand (the sharp point of a tack, for example) calls for extraordinary measures—just in case it turns out to be an actual threat.

Information (either contrast or novelty) is not alone in its ability to trigger arousal. The following stimulus characteristics also have arousal potential:

- *Biologic importance* (food, sex, noxious elements). Thus any allusion to food or sex by a product will contribute to its ability to arouse the viewer, especially if he or she hungers for either at the moment.
- *Intensity* (bright light, loud sound, strong smell, etc.) An intense yellow car will be more exciting than a pale, cream-colored one.
- *Certain spectral regions* (red light has more arousal potential than blue light; sounds in the 500- to 800-hertz region have less arousal potential than those on either side). Thus Apple managed to appeal to a range of tastes by offering iMacs in a range of colors. All other things being equal, a designer can make a blue product more exciting by changing its color to orange or red. Color alone can move a product from one end of the emotional-rational continuum toward the other. A white, black, or gray Bauhaus-inspired coffeemaker, residing at the rational, Teutonic end of the spectrum, moves dramatically to the emotional end—where more American and Italian offerings reside—just by painting it yellow, orange, or red.
- *Acquired importance* (a person's name, a recognized threat, an imagined threat such as a phobia, symbols or signs associated with important matters). It goes without saying that if you ran across a product in the store with your name on it—a person's name probably being the one word with the most acquired importance—you would get pretty excited.

Arousal involves activity in every cubic millimeter of the body. Some of its components (papillary dilation, for instance) go unnoticed by the aroused person. Many other components (such as changes in heart rate and regularity) are noticed, however. More subtle reactions are sensed, too, even if only subliminally. These include contractions of muscles in the walls of minute blood vessels near the skin and dilations of those in the muscles as the nervous system automatically shifts blood from the body's surface—where it is not needed and might leak through wounds—to muscles, where it can help to make the person more capable of fight or flight. Actions in all the organs of the viscera (heart, lungs, stomach, gut, etc.) are also sensed to some extent. Kinesthetic sensations, arising from movements—or preparations for movement—join the manifold.

Thus the perception arising from holding a tack between two fingers is felt throughout the body—and remembered to some extent throughout the rest of your life. A product's aesthetic impact—the extent to which it stimulates, excites, and motivates the viewer—corresponds to the amount of information it embodies, of course. Whatever the force of arousal, the whole manifold of sensations associated with it becomes part and parcel of the tack's identity. Furthermore, the sensations of this experience combine with those of previous encounters with tacks and similar things, which arousal now dredges up from memory.

Arousal evokes nothing less than the *whole-mind, whole-body experience* the psychologist and philosopher William James described as the stuff of emotions more than a century ago: "Our whole cubic capacity is sensibly alive; and each morsel of it contributes its pulsations of feeling, dim or sharp, pleasant, painful, or dubious, to that sense of personality that every one of us unfailingly carries with him." And, we might add, the "sense of personality" or daimon that every object carries with it.

The practical consequences of an increase in arousal are that you become temporarily smarter and stronger. Faced with the uncertainty of a novel stimulus, it prepares you for the worst that could happen by making you briefly stronger and smarter. You are thus better able to fight or flee—just in case the cause of the arousal represents a real threat. If it happens to be food on the hoof, it makes you a faster, more cunning predator. It doesn't matter that we already know the thing causing all the commotion is just a harmless product. We are conditioned to react that way to any surprise automatically. Arousal is important aesthetically because the rush and seep of sensations accompanying it account for much of the empathic character attached to a product.

CONATION AND ORIENTATION

Conation comes close on the heels of sensation and affect. Except for a listing in *The Dictionary of Psychology* for historical purposes, not one of the psychologists I asked could define the term, as none had ever encountered it. According to *The 1000 Most Obscure Words in the English Language,* "Conation is the area of one's active mentality that has to do with desire, volition, and striving." Conation is the most physical manifestation of awareness. Most of the feelings associated with the "shock of the new" arise from many visceral and skeletal reactions associated with the arousal mechanism. Conation is also proof positive that we are not only reactive creatures but also proactive predictors of the future.

Conation involves future-oriented volition, intention, striving—in general, urges to act. It is the seat of the will. You feel the intention, "I *will* do this or that," through bodily movements, actual or incipient, arising from kinesthetic sensory nerves buried in the muscles, joints, and tendons that inform you of bodily posture, weight, and movement. The kinesthetic sense lets you know, for example, the orientation of your arm, elbow, and hand even when you hold them out of sight behind your back. The will to squeeze the tack even tighter struggles with a growing urge to let go and escape the pain. At some level of discomfort, the urge wins out over the will. Conation gets bound up with the sensations and affects and also becomes a part of the unified memory of the experience.

An important part of conation is an unquenchable urge to look at the most pronounced source of information at any given moment. In the case of the tack, the gaze concentrates on the point, not the cap. This directed and concentrated attention is called the *orientation reaction.* Arousal enhances nerve transmission so that your eyes become keener than usual. So do your other senses. Orientation involves more than turning the head toward the stimulus and riveting the eyes on it; all your senses attend to it, even those such as hearing, taste, and smell that do not necessarily yield anything useful.

The nerves in your fingertip, where something significant is happening, become more sensitive. In *Psychology of the Arts,* Hans and Shulamith Kreitler call the orientation reaction "the physiological manifestation of the primary question: What is it?" It abets the nervous system's effort to glean as much information from the stimulus as possible—by throwing all available resources at it—in order to "make sense" of it as quickly as possible. Arousal remains keyed up and orientation locked on until redeeming answers are found.

You also feel compelled to look at the pointed end of the tack, where the information and sensation are concentrated, and disregard the insignificant cap. You focus your attention on the point, where information and thus sensation, conation, and affect are most concentrated and pronounced. Because conation has so much to do with posture and the subtlest of bodily gestures and expressions, conation is the prime source of body language and the clearest window into a person's otherwise inaccessible state of mind. Conative expressions, mimicked in a product's form, become the prime source of its empathic expression and the clearest view of its daimon.

COGNITION AND AUTHENTICITY

This crown jewel in the evolution of the human nervous system involves everything we ordinarily regard as thinking: pondering, imagining, contemplating, deliberating, reasoning, and judging. You are primarily in a cognitive mode as you read these words, ponder their meanings, and judge their worth.

A continuous stream of thought courses through your mind as you attempt to identify the stimulus, categorize it, decide whether it is good or bad for you, and otherwise "make sense" of it. As you squeeze the tack harder and the sensation at the point increases, it becomes more difficult to think of anything other than the point. Your thoughts on the matter run the gamut, "That's interesting," "It's beginning to hurt," "Why am I doing this?" "I shouldn't squeeze any harder, or I'll hurt myself," "Maybe, just a little more," "Ouch!" "What a dumb thing to do!" "Where did I put the Band-Aids?"

Cognition serves at least two purposes. First and foremost, it must determine that the novel stimulus causing the commotion represents no actual threat—or that it does. It also must overcome the discomfort of contradictory conclusions that psychologists call *cognitive dissonance.* Fundamentally, the mind cannot tolerate two contradictory notions plugged into the same categorical pigeonhole. It must bring them into agreement by modifying one—as in the case of a perceptual illusion—rejecting one, or shunning the stimulus altogether for lack of *authenticity.* In the final analysis, cognitive effort in the perception of a product involves a search for authenticity, a sense that it is genuine, reliable, and worthy of trust.

THE UNITY OF MIND AND BODY IN AWARENESS

Traditionally, psychologists treated sensation, affect, conation, and cognition as distinct and relatively independent states of mind—as if a person could think without feeling or willing or that emotions could exist alone with no thoughtful influence or any urge to act on them. The distinctions probably arose as a natural consequence of the philosophy of *dualism*—that separated mind from body, thinking from feeling, and ideas or ideals from realities—often associated with the likes of seventeenth-century French philosopher Rene Descartes. However, the notion of separate states of awareness flows from the very wellspring of most Western culture, Greece's golden age, some 20 centuries earlier.

Impressions that thoughts, emotions, and urges originate in different places perpetuate the notion of separateness. Cognition seems concentrated "up there" in the head's logical and rational *word brain,* to use Paul McLean's term. Affect seems to emanate more diffusely from "down there" in the torso and gut, from what McLean called the *visceral brain.* Conation, with sensations originating in the muscles, joints, and tendons from throughout the body as it acts or prepares to act, seems to inhabit and possess the entire body; we might call conation the *body sense.*

To some extent, the mental boundaries appear to fit the facts of the brain's structure. The tiny, almond-shaped amygdala near the center of the brain is now known to be largely responsible for what we experience as emotions, especially fear. And the brain's outer bark, the cortex, has long been known as the site of cognition. The boundaries are fuzzy, however. The olfactory lobe, sometimes called the *smell brain* because of its central role in taste and smell, also plays such an important role in affect that it also has been called the *emotional brain.* Consequently, odors are among the most emotionally charged stimulants. The hypothalamus, another primitive region of the brain responsible for automatically regulating body temperature and other essential life-support systems, also has been implicated in affect. Nor should we be entirely certain of the cortex's exclusive dominion over cognition. Because it has so many interconnections with other structures below it, how can we be certain that it is exclusively responsible for such things as thoughts and images?

In the final analysis, the manifold of impressions we call the *mind* constitutes a whole-body phenomenon. Everyday experience verifies that awareness is a blend of sensations, thoughts, feelings, appraisals, and urges. Thoughts always stir up feelings, such as when one worries about paying the bills. And sensations always dredge up memories and emotions associated with them. Smelling a gardenia might flood the mind with memories of the senior prom, decades ago, and the effects of bubbling adolescent hormones.

And all experience involves conation. When you observe a picture hanging slightly askew, you have an urge to walk over and straighten it. When I read after a day of writing at my keyboard, I often "feel" the words through incipient motions of my fingers, as though I were typing them. Conation has been shown to arouse affect and cognition, too. Willfully putting a smile on your face stimulates more pleasant thoughts and feelings than a scowl.

This integration of awareness is of tremendous benefit. It is not so important to be aware that the light is changing or that the sound is changing but rather that *something* out there is changing. That something may happen to be associated with changes in several manifestations of energy capable of affecting the organism—heat, light, pressure, etc.—so it is useful to have awareness of them all in a bunch. The fact that we fixate on the notion of separate *senses*—or states of awareness—has to do more with the nature of cognitive processes we are stuck with: We are compelled to *analyze* (divide perceptions into parts), *abstract* (pull certain parts out of a disassembled whole), and *categorize* (give the parts names).

It is important to realize, as James Gibson points out in *The Senses as Perceptual Systems,* that the primary importance of the senses is the gathering of information about the environment and the creation of percepts about what is out there—not the individual and particular sights, sounds, smells, tastes, and tactual sensations it evokes. The sensory systems are neither independent of each other, nor are they passive in their information-gathering tasks. He argues that

> The perceptual systems . . . correspond to the organs of active attention with which the organism is equipped. These bear some resemblance to the commonly recognized sense organs, but they differ in not being anatomical units capable of being dissected out of the body. Each perceptual system orients itself in appropriate ways for the pickup of environmental information, and depends on the general orientation system of the body. Head movements, ear movements, hand movements, nose and mouth movements, and eye movements are part and parcel of the perceptual systems they serve. These adjustments constitute modes of attention, . . . and they are senses only as the man in the street uses the term, not as the psychologist does. *They serve to explore the information available in sound, mechanical contact, chemical contact, and light.* [emphasis added]

From a practical point of view, it could hardly be any other way. Imagine the impracticality of a system where the specialized sensory receptors transmitted the particular forms of energy to which they were sensitive rather than transducing the energy into a common type of energy for transmission to the brain. The eyes might use fiberoptic conductors made of glass; they could transmit much more information to the brain each nanosecond than the relatively restricted and slow nerves they actually use. The ears might use hollow metal or glass tubes to transmit sounds directly and more efficiently to the brain. Touch receptors on the skin might employ hollow tubes filled with some incompressible liquid (water would do; pressure applied at the skin would be registered almost immediately at the brain as pressure). Heat sensors in the skin might be connected to the brain via highly conductive aluminum, silver, or gold. But they would be far too slow to protect us from burning ourselves. By the time the brain felt the heat, the finger would be well done. And these nerves would need very effective, probably very bulky, insulation to shield them from the body's internal heat.

In contrast, the electrochemical form of nerve conduction that has evolved is relatively easy to insulate from spurious electrical or chemical influences, and transmission is rapid and accurate—if not as efficacious—as the optical glass fiber that transmits data bits at the speed of light. The transmission of a signal along a nerve proceeds as a series of pulses that travel more like the baton passed from runner to runner in a relay race, not like an electric current along a wire that grows weaker the farther it goes and eventually dies out. The pulses are easily coded and boosted by additional releases of energy, stored throughout the length of the nervous path, so that the signals are identical to the signal at its source regardless of the number of times they have been divided at branches.

Nerve conduction is superior to ordinary electrical conduction, or any of the others described, in the same way that a bucket brigade is superior to a fire hose. It is slower than the hose, but it is more reliable if the fire is some distance from the hydrant or if there are several fires and only one hydrant. The signals leaving each branch can have the same strength as the single incoming signal. Thus an impulse originating at a single point somewhere in or on the organism can spread virtually unaltered throughout the whole organism, arriving at each point of destination with the same strength it had at its origin.

The amount of energy released by the sensory nerves is small but easily and efficiently transmitted to the remotest parts of the organism by the unique process of nerve conduction that resembles, but is not exactly like, electrical conduction. The regulatory mechanism of the autonomic system responds to this intermediate "nervous energy," not to the original forms of energy emanating from the real environment. Consequently, the sensory image of the environment is a derivation that equalizes all forms of environmental energy. All stimuli, regardless of their origins, have potentially the same importance as far as the organism's survival is concerned. Bodily contact, mechanically sensed through the skin, produces the same kind of neural event as a distant visual event. Under certain conditions, they will be identical, even though much less energy is involved in visual perceptions than in perceptions of pressure on the skin. If this were not so, we would tend to respond much more strongly to stimuli of touch and taste than we would to stimuli of sound, sight, and smell, regardless of the relative importance of the events with which they were associated. As a result, our actions would be related more to immediate events and less to probable future events. We might not take seriously an onrushing automobile that we perceive through the feeble force of light rays alone until it actually hits us with the compelling force and violence of physical contact.

From an embryologic perspective, a unified nervous system was the most practical and elegant way to go as well. Impulses end up much the same because they originated from identical cells that were, from the beginning, connected to each other to some degree. The entire nervous system starts out as the skin of the primitive slablike embryo. The skin, incidentally, ends up as the largest of the body's organs. Once the slab curls into a tube, the skin's cells begin to differentiate. Those ending up inside the tube develop into the brain and the cells lining the organs of the viscera. Those on the outside evolve into the specialized senses of the skin, sensitive to pressure, heat, and cold. Others become the most sensitive and important of all, the cells of the retina at the backs of the eyes.

THE PRIMACY OF SENSATION, AFFECT, AND CONATION

The different regions of the brain did not evolve simultaneously—or democratically—but by a process of accretion. Each evolutionary enhancement enveloped—and connected with—the older regions of the brain within or below it.

The deepest, more primitive regions of the brain, in both embryologic and evolutionary senses, take care of the most basic and essential requirements of life: temperature control, digestion, metabolism, breathing, and circulation, for example. They do this automatically, without bothering us with the need to think about these processes. In addition to being automatic, they are quick.

Not so the cortex, the most recent layer to evolve and develop, which literally caps off the brain (cortex comes from the Latin word for "the bark" of a tree). The deliberative cortex is slow—often agonizingly slow—when compared with the thoughtless but instant reflexes of less noble parts of the brain. It seems to have been contrived to delay reactions while it mulls things over: "Wait a moment," it cautions, "Let's not be too hasty. Let's think about what's going on and weigh the options before acting rashly." As marvelous as its cognitive skills of imagination, logic, rationality, and language are, the cortex is quite feeble, perhaps because, in evolutionary terms, it is so new and undeveloped. George A. Miller made its shortcomings evident in his classic paper of 1956, "The magical number seven, plus or minus two: some limits on our capacity for processing information." We can nominally hold only about *seven* pieces of information in short-term memory while preparing to process it. Thus local telephone numbers have only seven digits and, after unsuccessfully trying eleven-digit ZIP codes, the Postal Service settled on five or, for the extended version, nine. Even so, if someone interrupts you after you have looked up a number, with even a smidgen of new information, the number slips away.

Cognition bogs down readily because the cortex can process information only so fast; with regard to speed, it pales in comparison with even the cheapest palm-sized computer. The upshot is that while the cortex might seem in control of all it surveys from its commanding perch atop the nervous system—with all the latest neural equipment at its disposal—it is merely a figurehead when it comes to actually ruling the mind. If primacy of cognition is the definitive prerequisite of humanness, then we remain to this day more akin to beasts than that human ideal: Passion still rules, not reason. We *feel* the meanings of things more than we *think* them. This is why empathic communication exists in all things, even instances of linguistic and graphic communication.

It follows that feelings, not thoughts, are the final arbiters of action. This applies especially in matters of emergency, duress, and extreme stress. However, it also applies in all other matters, whether profound or mundane. When shopping for new pants, you check inside the waistband to be sure they match your waist and length (a cognitive matter). You could take along a tape measure and confirm even more measurements. You won't buy, however, until you've gone to the fitting room for confirmation by your senses; they must *feel* right and *look* right, regardless of what the labels claim. We never make a decision, regardless of how much thought we have given it, until it *feels right.*

Liking and Disliking

8

ALONG WITH A product's looks, the thoughts, feelings, urges, and appraisals associated with it become part of it in memory. They also meld with its associated stereotype to change it slightly or greatly, depending on the product's level of novelty. They also might shape the associated ideal by virtue of some disappointment: "Next time, they ought'a. . . ." These states of mind and body constitute the product's meaning and, from now on, the meaning of all other products associated with the same stereotype and ideal.

Of all these subjective associations, *appraisal*—an impression of relative goodness or badness—may contribute the most memorable part. In studies to determine the meanings associated with thousands of objects and concepts, Charles Osgood and his colleagues from every corner of the world found that fully half of all adjectives used to characterize them on their semantic differential surveys were evaluative in nature; *notions of good and bad outranked all others in every culture studied.* And none were more closely correlated with good and bad than *beautiful* and *ugly.* For all practical purposes, they are synonyms.

VALENCE: LIKING OR DISLIKING

Affective or cognitive appraisal correlates with conative *valence*—sensed as an urge to approach a good stimulus or avoid a bad one:

- *Positive valence* moves the viewer closer to the stimulus, for more intense sensations and fuller engagement of all senses—not just the distal senses of vision, hearing, and smell but also the proximal senses of touch and taste. In the ultimate case, the viewer wants to embrace or consume the stimulus.

- *Negative valence* moves the viewer to curtail contact by moving away from the stimulus or, at least, by turning the gaze away. In extreme cases, the viewer prepares to fight it or flee from it.

Liking something implies at least three states of mind and body. Bearing in mind that *love* is a synonym of *like,* the viewer derives pleasure or enjoyment in the object, wants to possess it, and in the deepest, most primordial sense, wants to become one with it by sharing its characteristics. One way to *share its characteristics* would be by being in physical contact with it or by imitating it. Museums have to put signs on displays asking viewers, "Please don't touch the art!" Hans and Shulamith Kreitler observed people contemplating Michelangelo's *David,* and within minutes, they were mimicking its posture.

Even the simplest, single-celled organisms display liking behavior. An amoeba moving aimlessly through its environment eventually bumps into something else. If it happens to be bigger and unlikable—as in something that likes amoebas for dinner—then the amoeba will back off, if possible. Usually, it is too late for the amoeba, and it will be enveloped by its predator, which demonstrates the most basic disadvantages of not having distal senses such as vision, for detecting things from a distance, and the predictive ability that goes with them. If the other thing happens to be smaller—and a source of amoeba nourishment—the amoeba becomes a consumer by enveloping and assimilating it; they become, quite literally, *alike.* If the amoeba had a mind and a voice, it might make a statement of *taste:* "Ah! That was *good!* I *liked* it. Indeed, I *loved* it!"

In loving someone or something—the ultimate example of liking—the one who loves wants to become one with the loved one. People *hopelessly* in love willingly *succumb* or *surrender* to each other. This is easy enough to understand in love of one person for another. Could it also be that someone feeling blue, perhaps unloved, shops for clothes in order to experience their embrace? Does this also explain guys in midlife crises, embraced by their sports cars?

GOLDILOCKS, THE GREEKS, AND THE INVERTED U-SHAPED CURVE

None of this discussion tells us basically why we like a product or dislike it. We still do not know where tastes come from or why they vary from person to person or even within the same person at different times. What causes a particular product to attract you but repel me? Why does your car (or house, or furniture) seem less (more) attractive now than when you bought it? Or why have they grown on you? Why do you like spinach and salads now but couldn't stand them as a child? What is the difference between beauty and ugliness?

The arousal and challenge of information, especially novelty, provides part of the answer. We like the challenge of learning. Why else would we read books, do crossword puzzles, and go to school after we do not have to anymore? We strive to overcome obstacles and opponents, weaving among social rules of the game, while trying to reach goals. The game is so little fun without the rules and would-be frustrations that we will create them to make the game more interesting.

Colin Martindale provides considerable evidence of an appetite for novelty—for its own sake. In *The Clockwork Muse,* he proposed that poetry and other forms of art must evolve through a series of innovative departures from established norms in order to maintain enough energy and vitality to hold their audiences. The evolution of styles, whether of poetry, music, or automobiles, depends on a succession of innovations to maintain interest and excitement. Skirts go up until they don't dare bare any more thigh. Then they go down until they don't dare to bare enough. Cars have skirts, too. Then they don't. Then they do again. Thus fashions go in cycles. Things that are old enough are new again.

Limits determine preferences, too. When Goldilocks sampled the Bear family's porridge, she found one bowl was still too hot, one was already too cold, but the third was "Just right!" The inverted U-shaped Goldilocks curve shows up time and again in studies of preference. The vertical *hedonic tone* axis of its graph represents how much something is liked. The horizontal axis represents a variable continuum, like those of a semantic profile: from cold to hot, from rational to emotional, from simple to complex. Few people prefer the

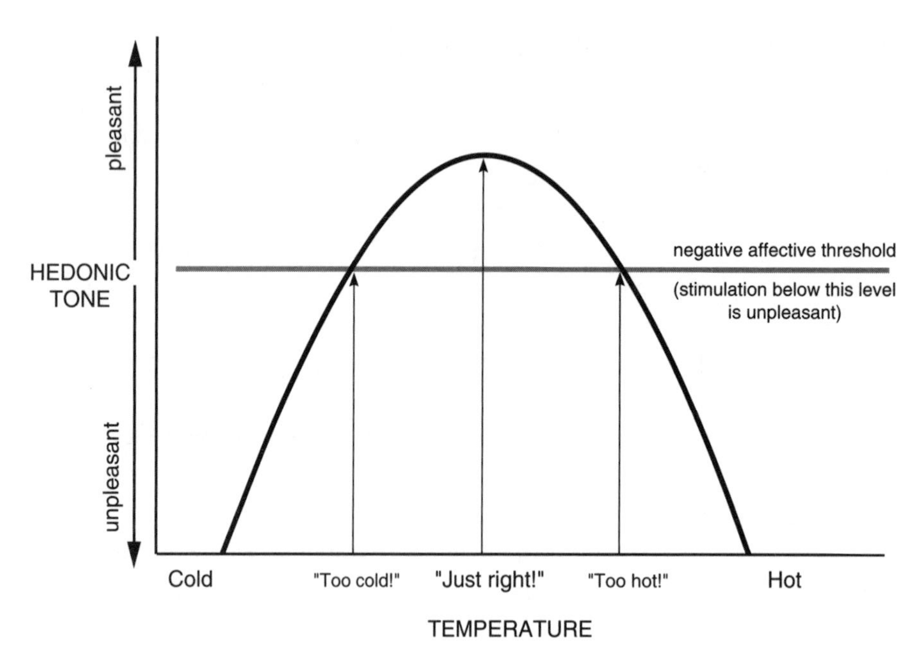

Goldilocks curve

most extreme characteristics, regardless of the attribute under consideration; they usually prefer some happy medium in between.

We can make the Goldilocks curve more general by labeling the temperature axis "arousal," "stimulation," or "information." "Temperature" itself is fairly general. Heat, a form of energy, also has been likened to information. Hot porridge embodies more information than less energetic cold porridge.

Wilhelm Wundt, the German psychologist who has been called the "founder of modern psychology"—and arguably the founder of psychological aesthetics—performed experiments over a century ago that gave objective substance to the Goldilocks principle. He found that *hedonic tone*—the relative pleasantness or unpleasantness of a stimulus—also peaked somewhere between extremes. A person who prefers sugar in his coffee won't like coffee without it, but neither will he like a cup with more than the amount of sugar he is used to—proving that you can have too much of a good thing. Starting with no sugar at all, he likes the coffee's taste more and more as the experimenter adds pinch after pinch of sugar. He continues to like it more until the sugar reaches the concentration he normally prefers. He then begins to like it progressively less as the experimenter continues to add sugar. Finally, he takes an intense dislike to the coffee. Indeed, some people have no happy medium; they like their coffee best with no sugar at all and quickly come to despise the coffee with the slightest addition of sugar.

Most people prefer lines, curves, and surfaces of relatively low degree (or complexity), but not the lowest. Some people prefer relatively simple figures formed of circles, parabolas, ellipses, and hyperbolas (second-degree curves). Others prefer the greater complexity found in fourth-degree curves. On average, however, studies show that most people prefer the gently undulating lines and surfaces of third-degree geometry—in other words, curves that are complex but not *too* complex. In one study, engineering students tended to prefer more highly organized second-degree curves, whereas music students tended to prefer more complex fourth-degree curves.

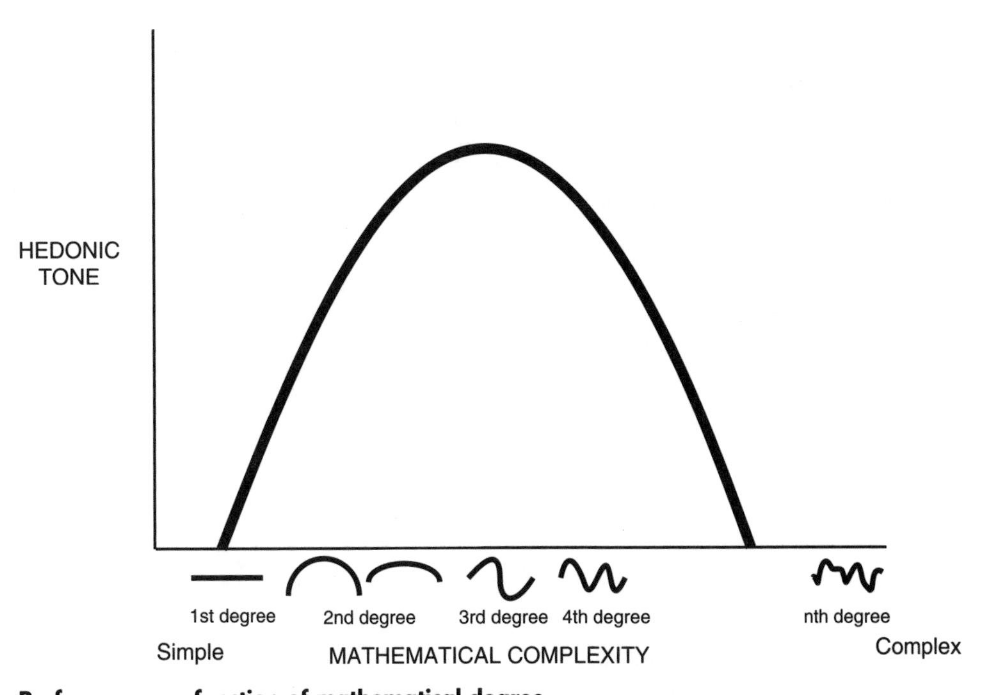

Preference as a function of mathematical degree

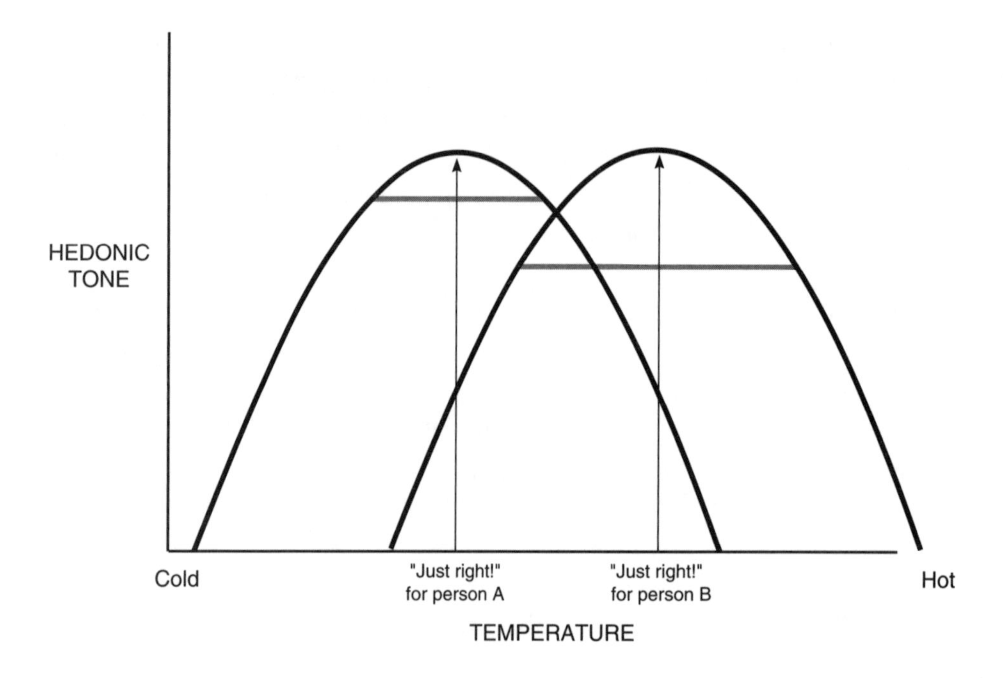

Just where preference peaks depends a lot on experience and what a person's used to. Thus, not everyone would have the same tastes for porridge as Goldilocks: In this graph, person A would prefer it cooler; person B would prefer it hotter. Their taste differences—what seems normal to them—might have been established when they were children, depending on Mom's cooking style. In other words, their porridge stereotypes differ. The tastes of persons A and B might have remained constant or changed over their lifetimes. Say person A went away to college and had a roommate who did all the cooking but served the porridge hotter than she was used to. If person A, not wanting to assume the cooking duties, kept quiet and got used to the new mode, she might well end up with a porridge stereotype hotter than the one with which she came to college.

Donald Hebb postulated an inverted U-shaped relationship between arousal and *cue function*—the ability to discriminate and learn from an experience. Optimal cue function—where the individual feels "bright-eyed and bushy-tailed" and performs at his or her best, mentally and physically—corresponds to what Hebb called the person's *optimal level of arousal* (OLA).

Our bodies depend on some basal level of arousal just to maintain circulation, respiration, and other essential aspects of metabolism. So arousal never sleeps, even when we do—although it is normally at its lowest then. It rises steeply as soon as you arise and accounts for increasing alertness and learning ability as you become more wide awake.

Since the optimal performance and learning ability associated with the OLA would serve us best under all circumstances, we might expect evolutionary pressures to promote behavior that maintains the OLA as often as possible. This is exactly what seems to happen. For as long as you remain conscious, you act in ways that would maintain the OLA. If the environment isn't interesting enough, you seek a more interesting one; if it's too noisy, you seek a quieter place.

If arousal exceeds a certain level for an extended period, behavior shifts to an aversive mode called the *avoidance reaction* or the *fight or flight response.* Although physiologically similar to the orientation reaction, it differs in significant ways aimed at a defense rather than inquisitiveness:

- The extremities become less sensitive instead of more sensitive.
- The pupils constrict.

If arousal continues to increase, it eventually reaches what Hebb called a *negative affective threshold.* Performance deteriorates, and the person experiences increasing agitation, emotional disturbance, and both cognitive and physical dysfunction. If it goes high enough and stays there long enough, complete behavioral disorganization can ensue. This probably accounts for why some people faint when faced with extremely high stimulation; losing consciousness amounts to the ultimate tactic for quickly escaping a stimulus and restoring a more normal arousal level. Primitive mammals, such as shrews and possums, who "play dead" in the face of even moderate threats probably are not acting; their negative affective thresholds probably lie so close to their OLAs that they faint dead away at the slightest provocation.

D. E. Berlyne, who led the resurrection of psychological aesthetics during the 1960s proposed that the low levels of stimulation associated with boredom were as stressful and distasteful as abnormally high levels—and actually were arousing, not restful, as we might assume. Exciting dreams might take care of maintaining the OLA while we sleep, when otherwise arousal might fall too far below the OLA. To the extent that we cannot find enough excitement in the external world while we are awake, we can make up the deficit from our imaginary world by daydreaming or, in extreme cases, hallucinating. Studies of sensory deprivation during the 1950s and 1960s verified this. After relatively brief periods floating in a swimming pool with body-temperature water, subjects wearing frosted goggles and earplugs that deprived them of virtually all sensory input, would begin hallucinating after a relatively short time. The hallucinations seemed so real that most subjects were terrified enough to bail out of the experiment.

Everyone has his or her own OLA and negative affective threshold. Some, with high levels, like it hot, active, and noisy. Those with low levels like it serene, simple, and quiet. Levels often change with age; I suspect that teenagers have characteristically higher levels than I do. They also change with circumstance. Someone moving from the country to the city might not be able to sleep well for some time. Eventually the person's OLA adjusts to the situation, and he has trouble sleeping when he goes back to the quiet farm to visit relatives. The contrary is just as likely; the quiet of the country can disturb the city dweller.

Arousal Jag and Arousal Boost

Berlyne observed that we enjoy brief, pronounced excursions of arousal well above the OLA. He called the shortest of these *arousal jags.* Jokes, surprises, and thrillers produce jags. We ride roller coasters, jump into snow banks after steam baths (now that's contrast!), and watch movies that terrify us in order to enjoy them. Beautiful products probably produce slightly longer ones. He called these *arousal boosts.*

Berlyne proposed that likes and dislikes might depend on stimulation of certain hedonic centers in the brain, identified by J. Olds. Increasing arousal stimulated both the *primary reward system* and the *aversion system,* but the reward system responded more quickly than the aversion system, so the initial, if brief, feeling was good. Although the aversion system was more slug-

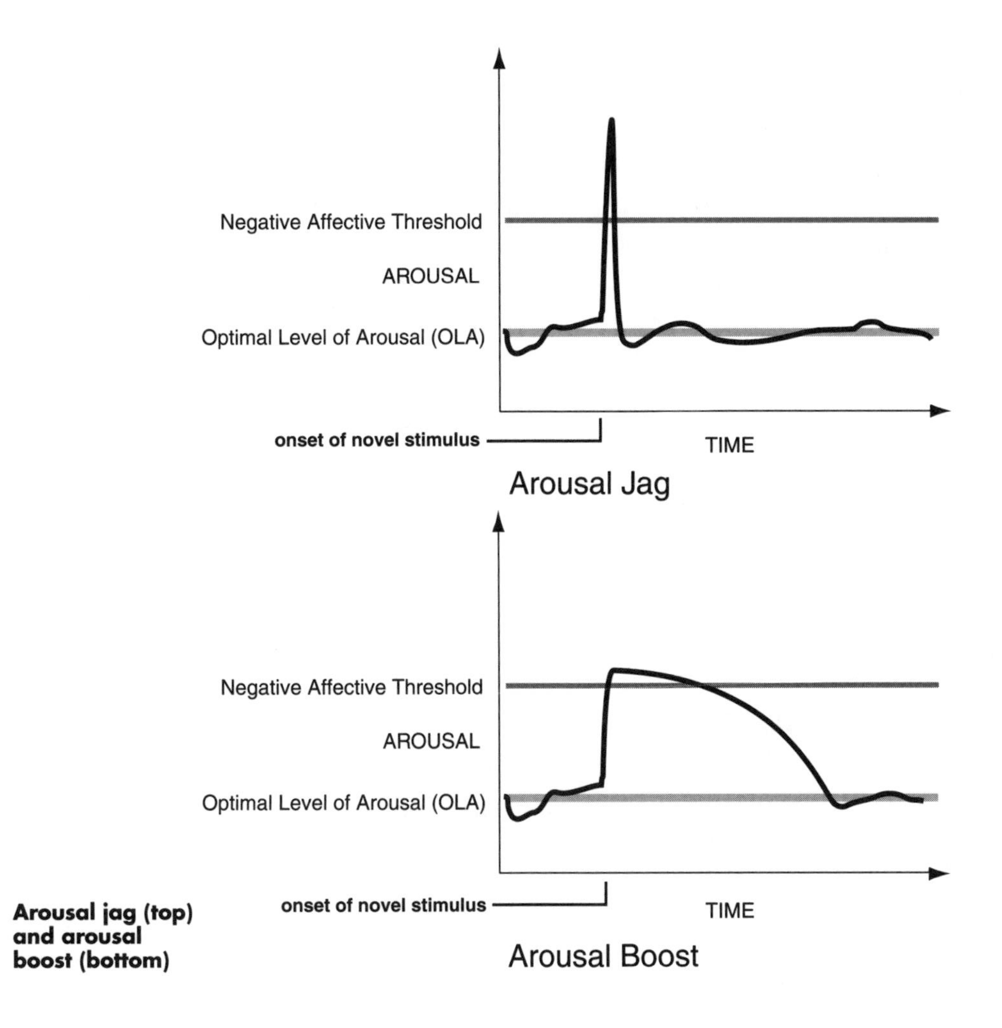

Arousal jag (top) and arousal boost (bottom)

gish in its response, it quickly gained the upper hand because, as it became more active, it sent inhibitory signals to the reward system that compounded its aversive affect. Unchecked, the hedonic tone quickly turned from good to bad. If the arousal continued to rise or even stayed at the same elevated level—such as when you don't get the punch line of a joke—the situation would get increasingly unlikable.

Only a decrease in arousal could check this pernicious pattern. As arousal subsided, the aversion system would shut down. This alone would have been sufficient to restore positive hedonic tone. But it felt especially good because arousal reductions yielded a hedonic bonus: Not only would the aversion system shut down, *a secondary* reward system would do its thing. The pleasure of sudden drops of arousal probably explains the sudden, sometimes euphoric elation of the "Aha!" experience felt on suddenly getting a punch line or solving a nagging problem. Getting the joke right away, though, brings the best of both sides of the arousal jag: the intense pleasure of the ride up, added to the equally intense pleasure of the ride down, all accomplished before the aversion system shakes itself awake.

Berlyne proposed that the area under the arousal curve corresponded to the psychological tension associated with a stimulus. The area must not exceed a certain value if the product is to be liked: *The higher the arousal curve, the shorter the period of arousal must be* if we are to enjoy the experience. An arousal jag can be tall because it is brief. An arousal boost cannot be as tall because it is more drawn out.

THE AFFECTIVE-COGNITIVE LAG

Joseph LeDoux's studies of brain functions suggest that affect and cognition proceed along two relatively distinct but interconnected pathways:

- The very *rapid affective system* is mediated largely by the *amygdala,* the small, almond-shaped structure of the brain's limbic system which is closely involved in emotion and motivation (conation). It goads the autonomic nervous system into action and spurs arousal. It also evokes emotional

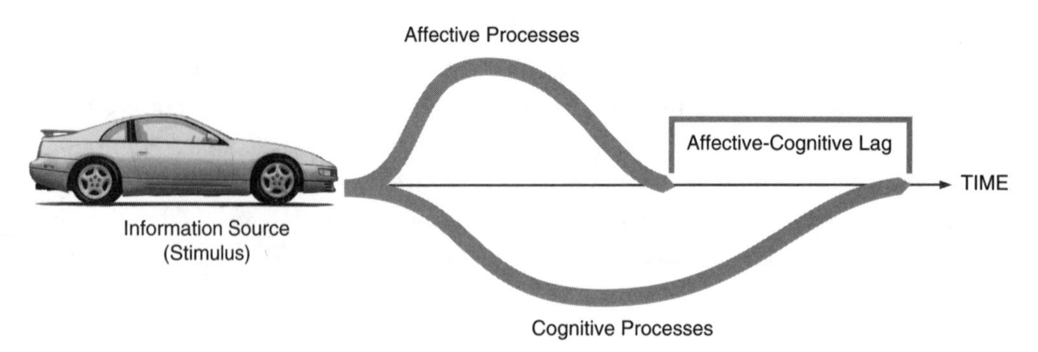

Affective-cognitive lag

memories associated with the stimulus. Affect entails an instantaneous "quick and dirty" appraisal of the stimulus, usually defaulting to a negative one—just to be on the safe side—unless it has reason to do otherwise.

- The much *slower cognitive system* is mediated largely by another part of the limbic system, the *hippocampus,* a seahorse-shaped structure. It is also in a position to influence emotion, motivation, and memory. But it has many more connections to the cortex (the avowed seat of cognition) than does the amygdala.

While affective and cognitive processes initiated by a stimulus begin simultaneously, they never finish together. Slow cognition, trying to make sense of the stimulus, always lags behind hair-trigger affect.

Regardless of how exciting a product's appearance is, whether we like or dislike it depends on how quickly we can cognitively make sense of its appearance. We enjoy the brief lags (corresponding to arousal jags), even very intense ones. And we tolerate moderately long lags (which correspond to arousal boosts). However, as the lag grows longer, even moderate boosts of arousal are more likely to agitate than to please. Beyond some length, they produce anxiety, loathing, and even fear.

Indeed, LeDoux's findings suggest that all emotions begin with fear, even if for only milliseconds; fear would seem to be the prototypical emotion. "Better safe than sorry," the nervous system seems to say: "Assume the worst until cognition proves otherwise." Thus our response to an unusual product always begins as prototypical fear—the initial, instantaneous urge is to flee by turning away or to stay and fight it by hurling invectives at it by calling it "Ugly!" Only if we can quickly dispel the uncertainty by finding meaning in the novelty will the fear turn to affection and a desire to draw closer to the object and have prolonged contact with it. Ironically, we derive our ability to like a new product from our very tendency to hate it. It doesn't matter that novel products seldom turn out to be actual threats, as novel things often did in our primitive ancestral past. Again, we take advantage of an *affordance*—instincts that emerged to serve one purpose, survival, coincidentally *afford* us the opportunity to enjoy beauty—as a fortuitous accident of evolution. We pay the price, however, of having to suffer ugliness as well.

Presumably, if our brains had evolved differently—and cognition proceeded as quickly as affect, permitting no affective-cognitive lag—we would like the looks of all products equally well, from the plainest to the most bizarre. *All products would be equally beautiful.* Or we might have no concept of beauty at all—nor of ugliness. In either case, our world would be a far less interesting place.

As matters stand, however, cognition always does lag behind affect. The only thing a designer can control is *how far behind.* This turns out to be a matter of how much concinnity a product has. Concinnity shortens the affective-cognitive lag by making information easier to process and thus accelerating the cognitive process. In the final analysis, then, two factors determine a product's aesthetic intensity and appeal:

- Its *arousal potential* (how exciting and interesting it is) is determined by its inherent information (some combination of contrast and novelty).
- Its *valence* (whether it is attractive or repulsive) is determined by its inherent concinnity (which determines the brevity of the affective-cognitive lag).

The next two chapters explain the two kinds of concinnity that designers need to master:

- *Objective concinnity* complements the objective form of information, contrast. Whereas contrast involves differences among a product's objectively observable elements, objective concinnity involves similarities among them.
- *Subjective concinnity* complements the subjective form of information, novelty. Whereas novelty involves differences between a product and imaginary entities such as stereotypes and ideals, subjective concinnity involves similarities between them.

Objective Concinnity

9

THE SPHERE HAS the maximum amount of objective concinnity possible in a three-dimensional form. Its counterpart, the circle, has the maximum amount possible in a two-dimensional form. Objective concinnity is largely responsible for the kind of beauty Plato liked: no perceptual difficulty, no emotional fuss. Museum curators like it, too, because of its timeless qualities. Objective concinnity, like contrast, can be readily expressed mathematically and in terms of bits and bytes. This means that a computer can be programmed to measure it, adjust it, and optimize it. It holds the greatest promise for computers with good taste. A product never loses or gains objective concinnity. A product will have the same amount at any time in the future as its designer gives it today. Products we think of as classics, those which show up in museum exhibitions and collections, tend to have a lot of it. If you like a product that seems to be "neat" or "clean"—or it just feels "right"—but you do not know why you like it, it probably has a lot of objective concinnity. The appeal of objective concinnity is so instinctive that it requires no thought. You know automatically that a picture hanging askew on the wall needs more objective concinnity, and you know instinctively just what to do to fix it. Applying concinnity, especially objective concinnity, is about doing what comes naturally; it's about *normalcy.*

Novelty—the kind of information designers usually rely on for packing a product with aesthetic potential—makes anything seem *less normal.* Adding concinnity—whether of the subjective or objective kind—to something with a lot of novelty restores some sense of normalcy by facilitating the cognitive processes that try to make sense of the abnormality.

Looking over the shoulder of an accomplished industrial designer, you will notice that once he has established the product's empathic themes and essential aspects of its form, he spends most of his time adjusting objective concinnity. He might simply line up elements that are almost aligned but not quite. Or he might be smoothing out curves that do not flow quite smoothly enough. Other theories of aesthetics have referred to notions similar to objective concinnity under the rubrics of *unity, simplicity,* or *symmetry.* The mathematician George D. Birkhoff called it *aesthetic measure.* The Gestalt psychologists called it *prägnanz* or, simply, *good form.* Objective concinnity comes closest to the original Roman definition of concinnity as an instance of *harmony, elegance,* or *symmetry.*

Most of the time, a designer strives to increase objective concinnity—especially during the tweaking, tuning, and refining of design's aesthetic end game. She does, in fact, pretty much what comes naturally to all of us. She fusses about the same things you do as you dress in the morning, taking care to choose just the right combination of items, just the right colors and textures, and just the right watch if you have more than one to choose from. Chances are you won't wear something that is obviously soiled, threadbare, or marred. You won't wear socks of different colors or one brown shoe and one black shoe. When you walk past the lawn to your car, you notice that the lawn needs mowing and the shrubs need trimming—two more instances of your instinctive need for objective concinnity. And while you're at it, you'd better wash the car (another instance). And you really should have that little dent taken out of the door (yet another). Note that none of these *strictly aesthetic* matters would really have mattered to Robinson Crusoe's ability to cope until Friday came along. We might have the natural urge to mind our concinnities, but sociocultural realities make it necessary. There is nothing quite so condemning as not minding your concinnities. Just try wearing different colored shoes to work. You might as well pick your nose.

Indeed, the principles seem rather obvious when you think about it—and remarkably easy to apply. In fact, anyone interested in becoming a designer would discover that learning the principles of objective concinnity and applying them are the easiest part of the job; they just need to follow their instinct for neatness. And objective concinnity delivers a lot of bang for the buck. Anyone who learns the list of general principles that will follow will have the wherewithal to improve virtually any product's design simply by increasing its objective concinnity—because hardly any designs have enough. As my beginning design students prove over and over, almost any design can be improved with an extra dose.

Essentially, objective concinnity runs counter to the influence of contrast, the objective form of information. Like contrast, the elements of a product to be manipulated are visible for anyone and everyone to see at the same time. Being so objective, they are susceptible to measurements and comparison. Contrary to contrast, objective concinnity is marked by *similarity, uniformity,* and *constancy* rather than by dissimilarity, variety, and variance. It speeds cognition and closes the affective-cognitive gap by reducing the mental effort required for

- *Analysis,* by *minimizing the number* of elements or attributes to be distinguished
- *Collation* and *abstraction,* by *minimizing the differences* among those which remain
- *Naming* or *categorization,* by providing elements that already have *familiar names* (straight line, sphere, etc.) and *don't require the effort to build new mental categories or stereotypes*

COGNITIVE HARDWARE

As we shall see later, subjective concinnity depends on the cognitive "software" programmed by experience. Objective concinnity depends chiefly on cognitive "hardware"—genetically determined neural structures—that provides instinctive notions of what is normal and requires no learning. Deciding to wear a brown outfit or a gray one depends on knowledge of current fashion and cultural norms, a matter of subjective concinnity. Deciding to wear socks of the same color is more instinctive, a matter of objective concinnity.

The instinctive nature of objective concinnity's effects on us probably is due to hardwired visual sensors that detect straight lines, vertical and horizontal lines, and angles of 45 and 90 degrees. It probably also stems from sensory modalities that develop earliest in the womb—such as touch and kinesthesis—and influence subsequent development of all the "later" senses like hearing and vision. The metaphors we use reflect how older modalities dominate newer modalities. A sound can be *sweet* or *sour,* for instance; taste is older than hearing. Words such as *light* and *dark* that refer particularly to vision are relatively rare and seldom refer downward to older senses. In a rare example, a wine taster might refer to the *light* or *dark* taste of a wine, but these terms are virtually synonymous with *light* and *heavy*—the kinesthetic notions he probably really means to evoke. It is much more common to refer to visual *tastes* in clothes, furniture, and other products. Most of our other visual metaphors draw on deeper sensory sources as well: A painting's composition is *balanced;* a sunset is *soft.*

Nature or Nurture?

We can't draw a sharp line between subjective and objective forms of concinnity because we can't draw a precise line between mental capacities and tendencies we are born with and those we learn; it is the old nature versus nurture problem that psychologists have always argued about and perhaps always will. Thus, many aspects of objective concinnity actually may be due to a sort of cognitive "firmware" instead of hardware. The sense of normalcy may begin in neural hardware we are born with but, nevertheless, remains malleable for at least some period after birth. It would become fixed and relatively rigid only after reinforcement—but still relatively universal in nature by virtue of the fact that we tend to share many of the same kinds of experiences in early life.

The fundamental normalcy of vertical, horizontal, and orthogonal characteristics provides a good example. These characteristics bombard our eyes as soon as we begin to see things as infants. The crib's bars are vertical. Its upper perimeter is horizontal. So is the mattress. The corners of the room beyond are vertical. The edges of the ceiling and floor are horizontal. Mommy and Daddy stand vertically (when they aren't bending over us). As you leave the crib to wander further afield, and throughout your entire life, the vertical and horizontal orientations and orthogonal relationships forever anchor your world view. Only rarely do the lines of a building or anything else built by humans depart from the basic "normal" orientations. Thus nurture could be as important in matters of objective concinnity as nature. Even so, it doesn't really matter for our purposes.

What matters is that vertical and horizontal orientations and orthogonal relationships—for whatever reasons—seem so normal to all of us that, for practical reasons, our preferences for them might as well have been formed and frozen prior to birth. They amount to cross-cultural constants. Horizontal, vertical, and orthogonal elements provoke so little affective disturbance and impose so little cognitive load—that they immediately make sense; there is no affective-cognitive lag to speak of. They require no contemplation, explanation, or justification. They just feel right.

The Sense of Normalcy

If there is a quintessential objective concinnity sense, it would be *kinesthesis,* the so-called body sense that is so important to conation and empathic communication. Kinesthesis literally means "feeling of motion," but it implies considerably more. It also has the unusual feature that, unlike vision, hearing, taste, smell, and touch, it has no direct connection to the outside world. It arises entirely from receptors inside the body called *interoceptors;* it is truly a body sense. It originates partly from nerve endings buried in muscles, tendons, and joints.

Of particular relevance to objective concinnity, it also arises from the *vestibular apparatus* in the labyrinths of the inner ears. Each ear contains three *semicircular canals* that lie in three different orthogonally related (perpendicular) planes. These canals amount to a gravity sensor that informs us about which way is up and from which we derive our notion of what is vertical. Fluid within them sloshes back and forth when the head moves in any direction. Fine, hairlike nerves wave with the flow like seaweed at the bottom of a tidal pool. They fire as they bend and send signals to the brain that the head is moving. They not only register tilts and rotations of the head in the three planes in which the canals lie, but they also register the associated accelerations and decelerations. The orthogonal arrangement of the canals probably explains why we naturally think and design in terms of three orthogonally related dimensions.

The normal state of the vestibular apparatus—when the signals from both ears are in equilibrium—occurs when the head is stationary and oriented vertically. Consequently, the eyes are aligned with the horizontal plane. In this orientation only, and when neither the head nor the body moves, the receptors in both ears register exactly the same *minimal-information* condition. This neutral state of equilibrium corresponds to the baseline for normalcy, perceptually and aesthetically. It is no accident that mathematicians refer to a line—generally, a vertical one—that stands perpendicular to another line or plane as a *normal.* Aesthetically, it is the least exciting, but it is the most comfortable and the most compatible.

The slightest tilt or movement of the head causes contrasting (and therefore informative) signals from the two ears that register as an abnormal situation—and a call for arousal. Thus begins the aesthetic experience—and the basis for future aesthetic experiences. A tilted or moving head becomes associated with a nonvertical, nonhorizontal, nonstatic—and abnormal—view of the world. It simultaneously evokes an elevated state of arousal and conative urges to restore comfortable equilibrium by adjusting the body in ways that would return the head to its normal upright position.

The visual and kinesthetic perceptions, along with the sensations of arousal, are inextricably bound together and stored in memory. Each similar experience thereafter recalls and reinforces them. And they receive plenty of reinforcement; your head and body are seldom in suspended animation. Because the entire manifold of sensations is bound together and remembered, your head doesn't actually have to move in order for you to have the associated aesthetic experience. All you need is movement of the scene across the retinas of your eyes. With no contradicting stimuli, the brain first assumes that the head, and perhaps the whole body, is moving instead of something in the environment. To make the subjective effect all the more compelling, it drags in all the kinesthetic and affective paraphernalia associated with such past experiences—perhaps even a touch of nausea—so that you subjectively think you are moving instead of some objective thing out there in the environment. *Empathy arises from this illusory subjective-objective bind.* If you have taken the roller coaster ride in an IMAX theater, you know how compelling the effect and affect can be. Or if you have mistakenly thought—and felt—your train pulling out of the station as you glimpsed the train on the next track start to move, you know how subtle the stimulus can be.

Actually, neither subject nor object has to move in order for the subject to experience empathy. Otherwise, we would not perceive motion in a parked sports car. Concentrations of information in a pointed nose and acutely sloped windshield do the trick by attracting the eyes toward the front end of the car and thus provide enough suggestion of motion to trigger the response. The tipping verticals and horizontals of a crooked picture are enough to evoke the incipient effort required to right the picture. Studies have shown that nerves leading to the viewer's leg muscles on the low side of a crooked picture fire more frequently than those on the high side—evidence enough that the subject feels the abnormal tilt.

GUIDELINES FOR INCREASING OBJECTIVE CONCINNITY

A product with a lot of objective concinnity will likely seem more stable, but also less dynamic than many other products. It probably will seem simpler and more unified, too. If it seems too stable, or too simple, or too plain, it probably has too much objective concinnity. In any case, because objective concinnity is literally *objective* in nature and right out there in plain sight, it is easy to identify by a few key hallmarks.

Orientation

- *Vertical* orientation of elements
- *Horizontal* orientation of elements
- *Orthogonal* (*right-angle*) relationships of elements

These are probably the most fundamental hallmarks, due to the importance of the vestibular component of the kinesthetic sense. They contribute importantly to impressions of uprightness, balance, correctness, and authenticity. Wherever I mention an *element* anywhere in this discussion, I refer to any noticeable feature or characteristic of a design that you can name or somehow measure: line, shape, proportion, surface, color, gloss, texture, etc. Obviously, geometric orientation has no relevance to color per se, but it could to a striped color pattern where the designer has the choice of orienting the stripes in any direction.

An important reason why we can interpret a watch with no numbers so readily is that it has four easily perceived cardinal positions that take advantage of vertical and horizontal relationships and orthogonality: at 12:00 and 6:00 (when both hands are aligned vertically) and at 3:00 and 9:00 (when the hands make right angles). Each hour corresponds to the straight-up orientation of the minute hand, and each half-hour corresponds to its straight-down orientation. The remaining eight "numbers" are easy to discern because the hour hand lies noticeably closer to one of the cardinal positions they fall between than the other. We can decipher those times precisely because of our superb sensitivity to vertical and orthogonal orientations. If the cardinal positions were rotated by just a few degrees, a numberless watch would be more difficult to tell time by. Note that no horizontal alignment of the hands ever occurs. The bottom pair of watches show the closest they ever come, at about 2:44 and 9:16.

Orthogonal relationships, also called *right* angles, constitute just one notion of *right*ness that creeps into descriptions and appraisals of objects marked by objective concinnity: normal, regular, correct, justified, orthogonal. A right angle formed by the junction of vertical and horizontal lines is the most objectively concinnous of all angles, but right angles at any orientation seem more normal and correct than any other angles. Hardwired structures in the eye and brain predispose us to perceive right angles. Right angles, like verticals and horizontals, impose virtually no cognitive load; a right angle always seems "right" right away. They are the quintessential angles by which we judge the relative perfection of all other angles.

If a corner of a *rect*angular (*rect-* as in cor*rect*) picture frame varies from 90 degrees even by a fraction of a degree, you can sense it immediately. And it offends your sense of propriety. Cognition lags increasingly farther behind affect as you continue to chew on it mentally, but unsuccessfully in your search for an appropriate reason for the unusual angle. You can only conclude that it is flawed, and your appraisal remains negative.

Often, though, a near miss is as good as a hit. It is best to adjust things until they *seem* right, not necessarily until they *are* right. Appearance matches reality in the case of a picture frame. However, I see many details that fall just short of perfect objective concinnity. This is especially true when lines aren't straight and surfaces aren't flat. Two curved lines that meet on the complex curved surface of a car might seem more "right" when they are not quite orthogonal. Designers should trust their intuition in these matters and go with it. I suspect that the preference for the imperfect objective concinnity has something to do with an instinctive desire to preserve the form's vitality by leaving a residue of mental tension.

Proportion

The following proportions, composed of all possible ratios of the integers from 1 to 5, seem especially appealing to most people. Proportions that on average get the most positive responses are darker in the figure:

- *1:1, 1:2, 1:3, 1:4, 1:5, 2:3, 2:5, 3:4, 3:5, and 4:5.*

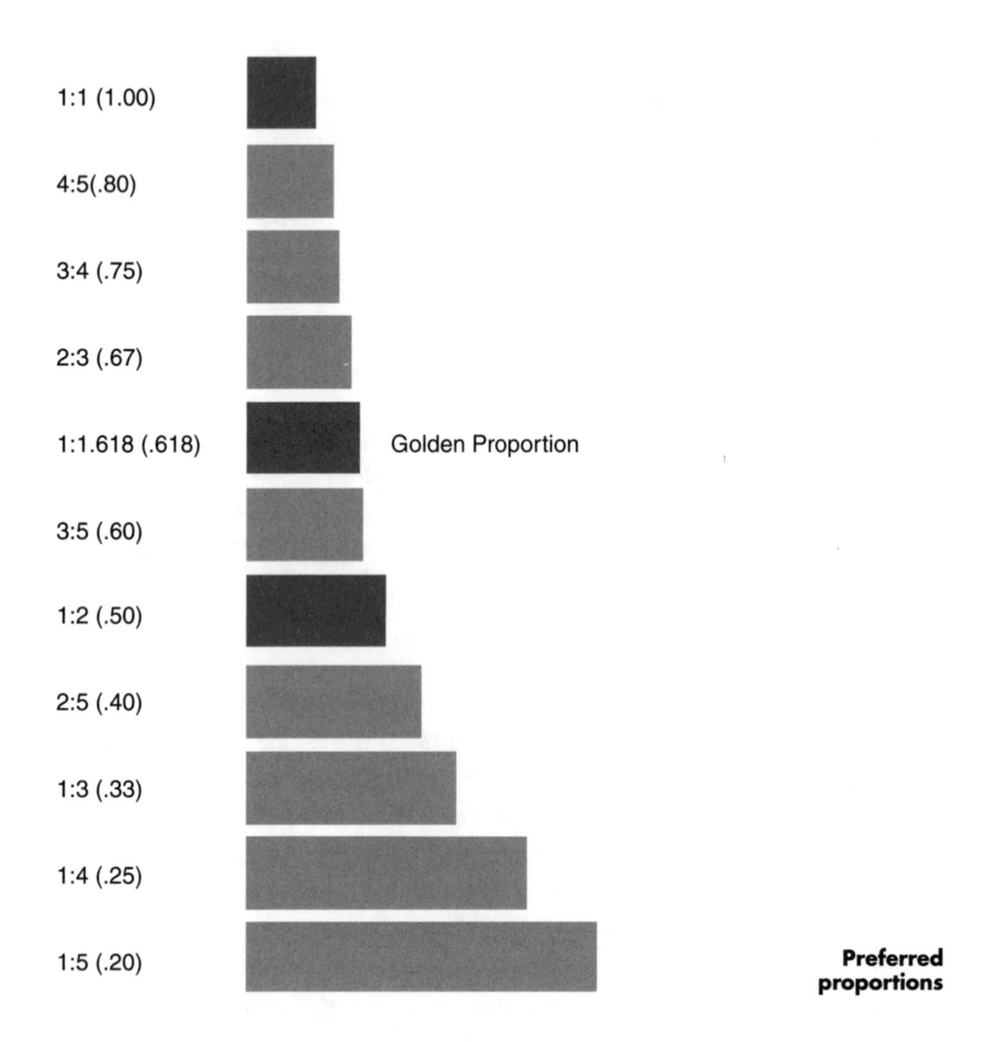

Preferred proportions

- *And 1:1.618,* the fabled *golden proportion* (also called the *golden mean, golden section,* or *divine proportion*), appears throughout nature. The Greeks (who named it) used it extensively, including the design of probably their most famous structure, the Parthenon. It is very close to 2:3 and 3:5. It displays an additional bit of mathematical magic: the reciprocal of 1.618 (1/1.618) is 0.618!

It may be impossible to know whether the golden proportion is a true case of objective concinnity. The universal preference for it suggests some inborn preference for it. However, what seems the most plausible explanation of its appeal (among probably thousands of theories) rouses the nature-nurture issue. It corresponds closely to the proportions of the human visual field. It might seem normal, and therefore appealing, simply because every scene we see throughout our lives is framed by the golden proportion. By extension, any object you see with the same proportion harmonizes with what has long since become the most normal proportion. If true, our preference for the golden proportion is a matter of mental programming based on experience—and therefore a matter of subjective concinnity—not neural wiring and objective concinnity. Nevertheless, the patterns are formed and reinforced from such an early age that we can regard it as given and a matter of objective concinnity.

Bear in mind that proportions apply to more than just rectangles. The ratio of any two comparable dimensions constitutes a proportion, including the lengths of two lines, the areas of two elements, the diameters of two circles, and the angles of two lines.

Continuity and Unity

The following hallmarks also apply:

- *Continuous* or *aligned* elements
- *Parallel* elements
- Alignment with *implicit landmarks*

An old aesthetic maxim states that good art always displays "variety within unity" (contrast) and "unity within variety" (objective concinnity). Unity reduces cognitive effort by reducing the apparent number of elements and thus the need for analysis, comparison, abstraction, and naming. In the ultimate case—when only one visually discernible element remains—cognition has nothing to take apart, nothing to distinguish, and no abstract generalizations to make. If that single element also fits readily into an existing stereotype—such as *vertical line*—cognition also has no naming to do.

Another important angle, 180 degrees, amounts to no angle at all, of course. Its legs lie in perfect *continuous alignment.* This brings up an issue of terminology that I need to deal with before specifying the next principle. Both continuity and alignment are important concepts that refer to the same thing: Alignment is a special case of continuity that involves only straight *lines;* continuity is the more general and powerful term—the one I want to use—because it also applies to curves. In the most general context, a straight line is technically a curve with zero curvature. It is what mathematicians call a *first-degree curve.* Circles, ellipses, and parabolas, which all curve in just one direction, are *second-degree curves.* Any curve that changes its direction of curvature must have at least *third-degree* status. Continuously joined curves or surfaces take on the same sense of unity as aligned straight elements. They seem to "flow" smoothly from one segment to the next, even when they are not joined but aligned across open spaces.

When elements do not actually touch, they should line up with obvious landmarks—such as centers of circles, endpoints or midpoints of lines, or tangents of curves. Simply make imaginary extensions of lines, curves, or surfaces in order to accomplish the alignment. They will seem unified even though they do not abut each other.

1996 Nissan 300 ZX

The 1990 to 1996 Nissan 300 ZX is an outstanding example of a design that is seminal in part because of its unusual objective concinnity. Until this car appeared, side windows were traditionally bounded by three distinct, curvilinear elements: the line formed by the windshield pillar (called the *A-pillar*), the C-pillar at the back, and a line joining them across the top of the windows. In the 300 ZX, the three elements have been integrated into a single, continuous curve arcing smoothly over the windows—a feature that has been widely emulated. This was accomplished partly by moving the windshield farther forward than usual, giving it what has come to be called a cab-forward look. Because the front edge of the door moved forward, too, the design made getting in and out much easier. In addition, the curve passes virtually through the wheel centers to create a symbolic, subjectively concinnous bridge structure that parallels the principal structure of the car. The line defining the bottom of the windows (called the *beltline*) is also continuous across the bottoms of the windows (almost a given) but also continuous with the line defining the rear hatch (another Nissan invention). This line, in turn, runs parallel to the deck-mounted wing of the 1996 model shown here and runs virtually parallel to an implicit line connecting the tops of the bumper panels.

The notion of continuity should be observed even when segments do not join physically, as in the "character" line that runs the length of the car, approximately through the wheel centers. It passes only implicitly through the body's wheel openings. It is important that this line seem uninterrupted because it forms a visual bridge with the curve over the windows. This visual bridge structure, tied together by the wheels, lends authenticity by paralleling the actual bridgelike structure under the skin that supports and stiffens the car.

The alignment of a watch's hands at 12:00 and 6:00 demonstrates just how powerful continuity can be: The eye can detect a discrepancy of as little as a second from perfect alignment. It is easy enough to understand why the discrepancy is so obvious around noon; assuming that the hands differ only in length, the hour hand becomes visible the instant the minute hand begins to uncover it. However, the misalignment just before or after 6:00 might not be as noticeable. At precisely 6:00, the aligned hands are wedded as a visual unit. A moment later this unitary line seems flawed as it breaks visually. Another moment later the unity evaporates as the line breaks into two separate lines.

The normalcy of continuity is especially appreciated when viewing a nerve-wracking case where elements are almost, but not quite continuous. It seems like a "fit and finish" problem, as though the elements were meant to be aligned, but someone in design, engineering, or assembly goofed. The result seems flawed or broken.

A designer working with unavoidable misalignment should exaggerate it enough that it seems *purposely* misaligned. Cognition will be greater than it would be with aligned elements but less than it would be if hung up on an unjustifiable flaw. By the same token, if colors of adjacent elements cannot be exactly the same, for a sense of unity, they should be decidedly different to avoid a sense of flawed unity. Consider, for example, early examples of body-colored plastic car bumpers that became discolored with age and drifted away from the body color.

As mentioned in Chapter 6, a curve or surface will seem kinked anywhere that its radius varies enough to exceed a certain perceptual threshold. The eye is able to notice the discontinuity of even a doubling of radius, especially on a polished surface, which amounts to just a one bit increase of information. Such surface breaks are acceptable, of course, where the designer wants a crease or bulge to appear. However, where the objective is a smoothly flowing surface that doesn't seem to break up into multiple pieces, the surface must meet certain mathematical conditions that prevent it from reaching the critical threshold.

The problem usually occurs, when using a CAD system, where the designer stitches together surface patches to create a product's overall surface. Without getting into too much detail, the patches have to be constructed of at least third-degree curves, which are also called *cubic curves.* Cubic curves are the only sure means for ensuring that two adjoining patches have both the same tangent and the same curvature at the joint. When tangent and curvature are the same, the combined curve is said to have *second-order* or *C2 continuity.* Simpler, second-degree curves, including circles, ellipses, and parabolas, only rarely achieve C2 continuity when joined. Two circular segments of different radii can be joined with the same tangent, to achieve first-order or C1 continuity, but the curvature at the joint will instantaneously change with the changing radii. Straight (first-degree) lines don't work at all, because they can't even be joined with the same tangent except in the trivial case of simply making a longer line; in all other cases they don't even have C1 continuity.

Recalling the discussion in Chapter 6, a curve's tangent is a first-order derivative, like velocity. Curvature is a second-order derivative, like acceleration, which is better at grabbing attention than first-order things. Generally, then, objective concinnity is greatest when curves and surface are constructed of third-degree curves. Traditionally, designers have accomplished this with drawing templates called *French curves* that maintain sufficient levels of continuity throughout curves of ever-changing radius. More recently, they have used CAD systems employing software for constructing *nonuniform rational B-splines* (NURBs), which are optimized for continuity. (Splines were long, flexible strips of wood, usually spruce, that designers of boats, airplanes, and cars formerly used to trace out long continuous curves.) Cubic curves are useful for another reason: They are the simplest curves, mathematically, that can change their direction of curvature to form S-shaped patterns. Fourth- and higher-degree curves permit ever-greater degrees of continuity, of course, and greater complexity.

The fact that CAD systems rely basically on cubic curves is aesthetically fortuitous because, as Berlyne reported, people tend to prefer them to simpler or more complex curves. They occupy that just-right region in the middle of the Goldilocks curve; they are complex enough to excite us, but simple enough to grasp quickly.

Ford Motor Company engineers learned the aesthetic importance of C2 continuity quite accidentally early in their exploration of computers as design tools. In the mid-1960s, when this happened, none of the company's computers were as powerful as today's desktop PCs. Thus, designing something as complex as the sculpted surface of a car's body quickly enough presented an enormous challenge. The simplest curves—circular, elliptical, and parabolic—would have used computer resources more efficiently than more complex cubic curves. However, aesthetics was not the issue. They were concerned about aerodynamics. Aircraft manufacturers and NASA had long since determined that bodies with the lowest air drag were characterized by third- or fourth-degree surfaces.

So they devised a wind tunnel test to find out just how serious the compromise of using second-degree surfaces would be. They built two all but identical models of a car design, one composed of second-degree surfaces and the other composed of third-degree surfaces. To no one's surprise, the third-degree model yielded the best results; a real car based on this design would have been more fuel efficient than the other. What did surprise them was that, eerily, the third-degree model somehow had a different character visually. It also *looked* better—even though the models seemed identical without careful examination. The eye seemed to be more disturbed by the subtle discontinuities of the second-degree model, just as the air was as it swept over the surface. The results provided yet another vindication of the classic aesthetic maxim that "form follows function."

Symmetry

Symmetry—literally "same measure"—is so commonplace in natural forms and human-made artifacts that we take it for granted. Synonymous with concinnity in many dictionaries, it has great power for increasing objective concinnity. Just one symmetrical replication of an element can make it seem normal and authentic, regardless of how unfamiliar or unexpected it might have been before. It actually belongs near the top of the list with vertical, horizontal, and orthogonal orientations. I left it until now so that I could discuss it in the context of principles I have already listed.

Max Factor, the Hollywood makeup legend, used a gridlike mask that resembled a catcher's face mask. When this mask was fitted over an actor's face, Factor could adjust the symmetry of the face, as well as any other characteristic that would increase objective concinnity to increase the person's appeal. But not too much. Some asymmetry, such as that provided by the mole on Marilyn Monroe's cheek, provides an extra measure of character and interest in the face—just as the asymmetry of an off-center nameplate does for a car's face.

There are more kinds of symmetry than the familiar bilateral symmetry of a car's face. A watch depends on radial symmetry that is accomplished by equal spaces among numbers displaced around its face. An object also can have more than one axis or plane of symmetry. And generally, increasing the number of axes or planes increases the objective concinnity. Whereas a watch has a single axis through the center of its face and a car has a single plane down its middle, a kaleidoscope has two, three, or more. Each additional plane increases the sense of unity. With enough planes, the randomness of a kaleidoscope's seed cell disappears in the unity of so many replications.

A vertical axis of symmetry compounds the effect of symmetry alone. Considering subjective concinnity, it yields the greatest sense of stability and normalcy. This is appropriate for a vase, which we expect to be stable, but not for a car, which we associate with mobility. Axes of any other orientation yield correspondingly more activity. Symmetry about a horizontal axis also implies stability, but without necessarily arresting sideways movement.

Miscellaneous Principles

- *Repeated elements.*

Symmetry is basically a form of replication constrained by the rules of symmetry. Generally, replication of any element makes it seem more normal; the least normal kind of element is a unique, one-of-a-kind element. Unique elements can look out of place unless they have special and obvious significance (subjective concinnity). The symmetry of the Corelle cup introduced in Chapter 6 is obvious. Its novel handle is appealing in part because of its objective concinnity. As seen in the illustration here, repeated circular elements

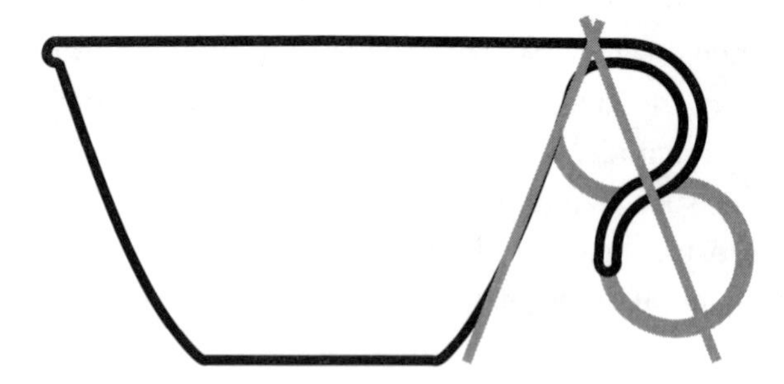

of the same size contribute to its objective concinnity, as does the symmetry between the line connecting the centers of the circular elements and the nominal slope of the cup's bowl.

In T. C. Chang's design for a saber saw, objective concinnity is rooted in the *repetition* and *regularity* of concentric circles and lines that radiate along incremental angles.

- Relatively *few elements* and a corollary, relatively *few actual or implicit intersections.*

The 1980 Cadillac Seville's objective concinnity depends on the fact that so many of its lines and curves, or their implicit extensions, intersect at common points. Elements *A, B,* and *C,* along with several other elements, all pass through the same intersection above the car. Other "landmarks" in the car's design, such as centers of wheels, also constitute intersections. Note how element *A* also passes through the center of the front wheel, a common feature of many car designs. Parallel lines and surfaces automatically minimize intersections because they never intersect. Neither do aligned elements. (Incidentally, elements *B* and *C* are also objectively concinnous because they are virtually symmetrical.)

Saber saw design

Reducing the number of actual and virtual intersections is ergonomically important because it can reduce distracting background clutter near sources of essential information and is also one of the most effective ways to improve a design's aesthetics.

- *Minimal variety* of elements.

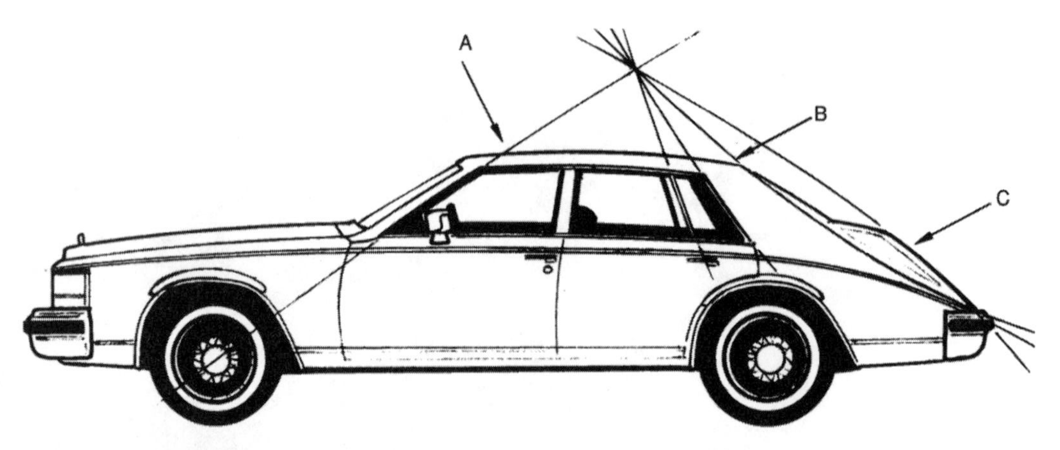

1980 Cadillac Seville

Lightolier desk lamp

- Where basically different kinds of elements do coexist, *bridge elements* contain attributes of the dissimilar elements that tie them together with a "family" resemblance.

A classic low-voltage desk lamp by Lightolier employs just two basic geometric elements, a spherical shade and a cube base. These are connected by a telescoping third element, a cylinder that bridges them not only mechanically but also geometrically. A horizontal cross section of the cylindrical bridge yields a circle (as any cross section of the spherical shade would); a vertical cross section yields a rectangle (as most sections through cubic base would).

Although 45-degree angles do not seem as normal as 90-degree angles, they do seem more normal than other acute angles. They have the additional attribute of adding some excitement where right angles might stultify a design. Thus we can add:

- Elements that do not intersect at 90-degree (orthogonal) angles often *intersect* at *45 degrees.*

The objective concinnity of 45-degree angles would make a watch with an 8-hour face easier to interpret than the 10-hour face the French introduced in the eighteenth century in an idealistic but ill-considered attempt to "go metric" with time and bring it into consistency with other scientific measures. They quickly abandoned it. Not only did it buck an old and familiar tradition, but it also would have been more difficult to read because it had only two objectively concinnous cardinal positions, at 12:00 and 5:00. An 8-hour face would have been a better option with 4 numbers in orthogonal positions and 4 at 45 degrees.

Anyone who uses a graphic program on a computer knows that there are a number of similar permutations that can be performed on all or part of an object: scaling, skewing, stretching, shrinking, rotating, etc. Generally, objective concinnity of the result usually is greatest when the same operation is applied to all elements of an object.

- Shapes are often derived from *simple permutations* of other, more objectively concinnous shapes.

Iron design

Chris Schmidt's iron design began as two simple cylinders joined by a simple rectilinear block. More complex and interesting aspects of its form arose when the profile of its base was rotated through the basic assembly like a knife—shrinking as it did so that the point of the base followed the predetermined curve of the iron's profile. Despite the simplicity of this compound permutation—executed with a computer—it was responsible for most of the iron's aesthetic character, including the subtle S-shaped line meandering over the surface along just the right path.

Symmetry doesn't always lead to practical form, of course, or necessarily attractive form. The situation often can be rectified without destroying symmetry's benefits by judiciously applying a relatively uniform transformation. By *uniform,* I mean consistent and simple, such as moving every point along a curve in the same direction or expanding or shrinking everything by the same amount.

COMPLEMENTARY VERSUS NEGATIVE INFORMATION

Since many of the preceding principles involve reduction of variance—variance being the essence of information—why don't I simply call objective concinnity *negative information?* This would simplify my task of explaining what's going on and simplify the designer's problem to one of merely adjusting the amount of information in a design: Increasing novelty or contrast would increase information; decreasing novelty or contrast would decrease it.

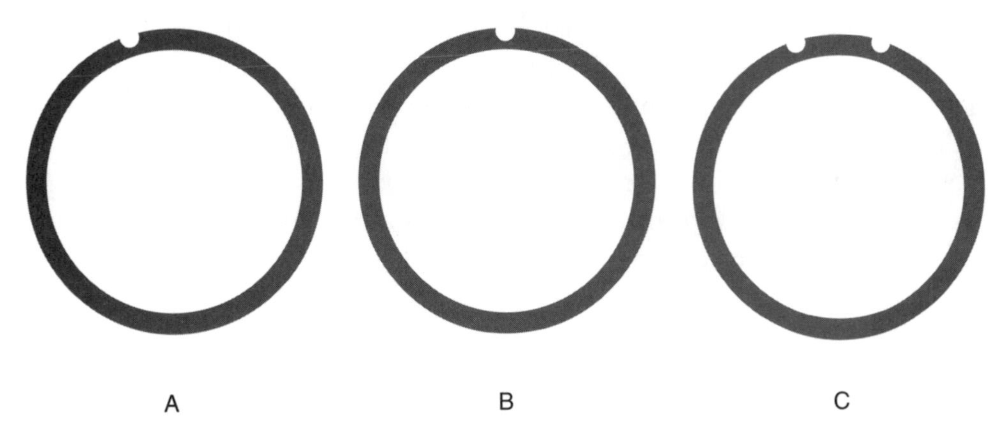

A B C

Objective concinnity is not simply the opposite of information because a designer can *increase* objective concinnity by either decreasing *information or increasing it.*

The depression in ring *A* (additional information) *deforms* it. This is tantamount to saying that it has no name of its own; it does not fit neatly into any preexisting categorical stereotype. To become acceptable would require additional, time-consuming cognitive effort to name it and establish a new stereotype (What would you name it? See what I mean?). We could restore the propriety of the figure simply by removing the depression (subtracting information). Or we could fix it by rotating the depression to the objectively concinnous 12:00 o'clock position, as in ring *B*. It would seem more normal and authentic—even without explanation of the depression's purpose. The depression in ring *B* seems more like the result of an intentional act, not an accident. It would fit comfortably with the "watch face" or "watch bezel" stereotype, for instance, and for obvious reasons. We also could increase objective concinnity, as in ring *C*, by adding a second depression, symmetrical with the first about the vertical centerline (additional information). As with ring *B*, it is highly unlikely that such a configuration would have occurred by accident. To be symmetrical, it must have been designed purposely. Even though this figure is as difficult to classify as ring *A*—it doesn't fit readily into an existing stereotype either—we are more inclined to give it the benefit of the doubt and assume that it is an authentic something—although we don't know what.

Thus, objective concinnity is not simply equivalent with negative information. Rather, it is the *complement* of information. Its virtue lies in its ability to complement (complete) the information process more expeditiously and satisfactorily—rather than in any ability to counter information's effects by canceling it out. Mixing information and concinnity is much like mixing what artists and color scientists call *complementary colors*—such as red and green. Adding green paint to red paint does not eliminate the red. It is still there technically,

Car with front-rear symmetry

but the mixture has a different and more neutral character. Theoretically, if we mixed equal amounts of red and green, we would end up with a neutral gray color, but without eliminating any of the red or green paint.

The gray is less interesting and exciting than red—or green. A design can have too much objective concinnity. While multiple kinds of symmetry seem quite natural and proper in a watch, only left-right symmetry seems appropriate in the design of a car. A car with front-rear symmetry stops dead in its tracks visually and loses much of its essential sense of action and movement. This brings us to the matter of how objective concinnity interacts with subjective concinnity.

INTERACTIONS WITH SUBJECTIVE CONCINNITY

Objective concinnity is almost a design panacea, but not quite. While we all hunger for it and search for it constantly, each of our appetites has a limit: Some like more; some like less. Probably no one has an appetite that exceeds Plato's. Although Plato did not use my terminology, he had objective concinnity in mind when he spoke of ideal form; Plato was an objective concinnity junkie. If he were in charge of a company's industrial design office, every product would emerge as some derivative of spherical form. Or if rolling off tables were a problem, he would do the next best thing and give it a cubic form.

Industrial designers would be ill-advised to follow Plato's single-minded lead by maximizing objective concinnity in every case. Not only would most products end up being impractical, ergonomic disasters, such as Plato's spherical TV set, they would likely be crashing bores aesthetically. The problem is not with objective concinnity per se but with how it affects subjective concinnity.

To understand the problem, consider this list of semantic differential scales:

rational–emotional
static–dynamic
calm–busy
smooth–rough
clean–dirty
organized–chaotic

unified–disparate
integrated–disintegrated
cold–hot
simple–complex
plain–ornate
neat–messy
passive–active
boring–interesting

Scores on these particular scales behave in predictable fashion as a product's objective concinnity changes. Any design change that increases objective concinnity—aligning elements, making them symmetrical, etc.—moves scores leftward on all scales. In other words, increasing a product's objective concinnity makes it seem more rational and less dynamic. Conversely, any design change that adds information—either contrast or novelty—moves scores to the right; the product will seem more emotionally dynamic. Thus, the designer can move the scores of a product's semantic profile closer to the scores of the ideal—thereby increasing its subjective concinnity—by carefully adjusting the product's objective concinnity and information.

The objective, of course, is to *maximize subjective concinnity* by making the product's semantic profile coincide as much as possible with that of the ideal. While you can have too much objective concinnity, you can never have too much subjective concinnity. The object in adjusting objective concinnity is seldom, if ever, to maximize it; the designer strives instead to *optimize objective concinnity*—only insofar as it serves to *maximize subjective concinnity.*

As you can see, a design with lots of objective concinnity—rational, static, calm, smooth, clean, etc.—will seem cold, simple, passive, and just possibly, boring—rather like modern design, straight out of the Bauhaus. If there was a guiding aesthetic principle at the Bauhaus, one that stood out above all others, it was objective concinnity. As I have suggested, products designed at the Bauhaus or inspired by it tended to take up residence near the rational end of the rational-emotional scale. Additional objective concinnity is one way to improve a design that seems too busy or emotional (or what is the same thing, not rational enough). If the product already seems too rational, it might be fixed simply by backing off on objective concinnity or increasing information with more contrast or novelty.

Extreme amounts of objective concinnity are appropriate for products expected to look plain, cold, even uninteresting (refrigerators come to mind here). Watch designs, each with appeals of its own, span the gamut; some have extreme amounts of objective concinnity, and others have relatively little.

In each case, the designer must aim for just the right degree of objective concinnity. Perhaps the chicken's aspherical egg—which Raymond Loewy singled out as so perfect that no designer could improve it—has just the right amount. An egg's form, like any other with precisely the right amount of objective concinnity, seems at a critical—perilous—pinnacle of perfection. Ignoring the material fragility of its shell, it seems that any attempt to revise its delicate form even slightly would upset some critical balance of tensions within the membrane and cause it to disintegrate like a soap bubble. The designer experiences supreme elation whenever he or she senses the moment when the design teeters on the razor's edge of insufficient objective concinnity or too much. Like hitting the "sweet spot" in tennis, golf, or baseball—or any other Zen moment—it amounts to a spiritual peak that keeps industrial designers coming back for more.

Subjective Concinnity

10

THE DOMINANT ARGUMENT in aesthetics reminds me of the one in physics concerning the nature of light: Does light consist of waves or particles? Both notions are valid, but they tend to be mutually exclusive. For any given situation, you can explain light in terms of waves or particles, but not both. A similar difficulty hampers aesthetics.

Two general definitions of beauty have coexisted for as long as people have argued about the nature of beauty. Objective concinnity pertains to the objective model of beauty, reminiscent of Plato. It holds that some ideal form exists for every object. Once the object acquires that one perfect form, everyone will like it. And it will remain beautiful for all time. You can best explain a timeless classic, such as the Horwitt watch, in terms of objective concinnity. The subjective model, which holds that "beauty is within the eye of the beholder," assumes that beauty is a matter of individual tastes and that no two people are destined to like exactly the same things. The appeal of an SUV can be explained best in terms of subjective concinnity. In the end, the Horwitt watch will always be revered by many. SUVs, however, like the finned cars of the 1950s, eventually will be reviled by many.

Critics, who contend that aesthetics cannot be rationalized, always cite the subjective model as proof. But they are wrong. Objective and subjective forms of concinnity, unlike waves and particles, are compatible and complementary. Thus the theory presented here really does permit beauty to be rationalized and, what's most important for product designers, made predictable.

Like objective concinnity, subjective concinnity makes things seem normal and appropriate—novelty having made them seem at first glance abnormal or weird. However, the underlying mechanisms differ. Each of us senses the objective concinnity of a thing and responds to it in much the same way, as if our response is due to hardwired circuits in the brain we are born with: If a car's bumpers are the same color as its body, it seems more unified *to everyone* than a car with contrasting bumpers of a different color. Not necessarily more beautiful, just more unified. If you like a sense of unity, this will help you to like it.

The objective concinnity inherent in the uprightness and symmetry of the Horwitt watch can be observed and measured directly. Because objectively concinnous properties do not diminish over time, the appeal of the Horwitt watch persists from decade to decade. This timeless character is an important reason why, in 1960, New York's Museum of Modern Art added it to its permanent collection.

Many other examples of concinnity cannot be seen or measured readily because they depend on a product's design to be in sync with subjective beliefs, values, expectations, and ideals. Because subjective matters differ from one person to another, different people find concinnity in different things. We all come to like different things. When your spouse thinks the easy chair you love is ugly, he or she is simply blind to the concinnity you perceive in it. The appeal of some products can be quite volatile because values, beliefs, and concerns change over time. And individual tastes change with them. When you come to like something you once thought ugly or dull, it has gained concinnity in the meantime not because it has somehow changed but because something inside your head has changed. By the same token, if you eventually come to loathe the design of that SUV you fell in love with last year, it probably will be because your values will have changed—and the SUV will have lost concinnity in the process.

The upshot is that designing a product that is subjectively concinnous for everyone forever is probably impossible. With respect to subjective concinnity, beauty *is* truly within the eye of the beholder. If you like a sense of unity in your automobiles, then body-colored bumpers will lend the car more subjective concinnity than contrasting bumpers. However, what seems appropriate to you will not necessarily seem appropriate to others. If you or someone else would rather have bumpers that stand out visually from the body—perhaps because they seem tougher and more protective—then you probably would prefer contrasting colors. Furthermore, there is no guarantee that you will like tomorrow what you like today—or what you hate today, you'll love next year. Times change, and along with them, personal values, tastes, and preoccupations change. People from other places and groups will not like what you like either, and vice versa, just because, for each of you, values, tastes, and preoccupations have always differed.

At any rate, subjective concinnity does present tougher problems to designers than good ol' measurable objective concinnity because we have to get inside people's heads—which are often murky and messy places—and measure things with more problematic instruments than rulers and light meters.

A Cup of Subjective Concinnity

Corning's Corelle teacup, with the "beaver tail" handle, already qualifies as a classic. It looks quite normal—except for its original handle—which differs markedly from the typical cup handle. Therein lies its aesthetic potential. Except for the handle, there would be nothing worthy of discussion or special consideration.

The handle is unusual enough—as revealed by comparing it with its stereotype—to grab a shopper's attention as she saunters the aisles of a china shop. Without giving the matter a thought, the shopper retrieves her cup stereotype from memory and compares it, lightening fast, with the cup before her. Information—novelty—has done its job; like the sharp end of a tack, it has gripped her attention, switched on her orientation reaction, and provoked an arousal boost.

Now the race is on. Affect jumps off to its usual quick lead. Now the question is: Will the shopper's cognitive skills make sense of the novel handle quickly enough to quell abnormally high or prolonged arousal—and let her like it? Or, perplexed by the weird form, will the shopper abandon it and go looking for a competitor's offering? Some concinnity would make it easier.

The shopper is excited by what she sees but not necessarily pleased; remember, ugly and threatening things are exciting, too. The shopper is having trouble finding reasons for the handle's unusual shape, based on her current knowledge.

The handle is not inherently unattractive because it obviously has considerable objective concinnity. It forms a pleasant, quite elegant S-curve. Just two cylindrical segments of the same radius comprise it. Nevertheless, the situation remains, for the moment at least, very much in doubt. The shopper's first instinctive thoughts are that something is wrong with this particular handle, that it is de*formed,* mal*formed,* or flawed. If this appraisal drags on, giving arousal time to build to an intolerable level, the negative valence might become unshakable enough that the shopper will turn away from it, put it out of her mind, and consider alternative products. Only to the extent that the shopper can cognitively rationalize the handle's strange form, in relatively short order, will the incipient disgust change to desire.

Thus it isn't a lack of objective concinnity that bothers her. It is a suspicion of inappropriateness of form that bothers her, and strictly a matter of insufficient subjective concinnity. The handle's form is obviously not going to change to conform to her expectations. So until something inside her head changes—until she *comes to believe* that the handle's "form follows some function," for instance—she can't justify it.

In an effort to find the logic and rationality in the handle's form, the shopper's cognitive faculties are racing to catch up with the affect it has triggered. If cognition can sort things out quickly enough to reverse the rising tide of arousal—turning it into just a momentary arousal boost—the shopper will be rewarded with good feelings and a positive valence toward the cup. How quickly cognition accomplishes its mission will determine whether she ends up disliking and dismissing the cup's design or—as the manufacturer fervently hopes—liking it enough to buy a set. Anything the designer can do to speed the cognitive process and close the affective-cognitive gap will help.

To make sense of the novel feature, cognitive effort must be spent on at least the following tasks:

- *Analysis,* whereby the viewer mentally breaks down the product's semblance into what seems to be its constituent parts, goes on continuously. The viewer is often aware of what she is doing, but analysis always occurs below the threshold of consciousness as well.

Even though the cup quite obviously consists of just one piece, it nevertheless *seems* to have at least two—the bowl and handle. They are separately discernible not only because the handle's form differs markedly from the bowl's but also because of the relatively small radii—and thus large concentrations of information—where it seems somebody joined the handle to its body. Actually, the concentrations of information are such that the cup readily breaks into four elements: the bowl, two partial cylinders comprising the handle, and a bead that runs around the top periphery of the bowl and continuously around the handle opening at the top of the bowl and the periphery of the handle.

- *Abstraction* or *generalization,* whereby the viewer literally "pulls out" properties of the product that have no concrete substance themselves but exist only by virtue of a "host" object, also goes on. Abstractions are like adjectives that cannot exist without the nouns they give character to. Color, for instance, is an abstract property. *Whiteness* is an accurate abstraction of the cup; all its elements—bowl, handle (partial cylinders), and bead—are white. So is *circularity;* making a horizontal slice through the bowl at any height reveals a circular cross section, and so do slices through the bead. In addition, the cylindrical elements of the handle also have circular cross sections.
- *Categorization,* where the viewer *names* the *product,* its *parts,* and its *abstract properties* by filing them, in effect, in mental pigeonholes, also goes on. The Corelle cup's handle eventually must fit into the pigeonhole labeled "cup handles" or, at least, "handles." As long as it remains in categorical limbo without vindication, the handle and, by association, the entire cup will continue to disturb the viewer.

Fundamentally, the designer increases concinnity by

- Decreasing the load of analysis by reducing the number of discernible elements in the product's form
- Decreasing the load of abstraction by reducing the number of possible abstractions
- Decreasing the load of categorizing or naming products or elements by reminding viewers of well-established, appropriate categories—rather than burdening them with the additional mental effort of creating new names for them

Subjective concinnity assists naming. Objective concinnity assists analysis and abstraction more than anything else. The fact that the cup as a whole has little information and ample objective concinnity reduces the cognitive effort of analysis, comparison, abstraction, and naming at the outset.

Analysis goes quickly because it has only three discernible parts: a radially symmetrical bowl with a subtly curved profile, a handle, and a cylindrical bead that runs around the edge of the bowl and handle to unify the whole assembly. The concave top surface of the handle flows smoothly into the concavity inside the bowl to enhance the unity of the assembly.

Further analysis reveals that the handle's S shape is composed of two tubular segments of the same diameter. And a line through their centers is essentially symmetrical with the slope of the cup's profile.

Abstraction is easier when the parts to be "pulled out" of the larger form can be named readily because they belong to well-defined categories, as in the case of the tubular elements making up the handle. Comparison is effortless because the two elements are symmetrical: The tubular elements literally have the "same measurements," the same diameter, and the same wall thickness. All elements—bowl, handle parts, and lip—are radially symmetrical. While it seems trivial to say so, all components are also the same color. These similarities make abstraction and comparison easier.

The handle's subjective concinnity emerges and does its job of reducing arousal as soon as the viewer begins to appreciate the practical and ergonomic advantages its unusual form brings to the cup.

- As a practical matter, the handle's open bottom allows eight cups (each of which is 2⅜ inches high) to be stacked and stored in a space only about 3 inches higher. Realizing that it satisfies so well the functional requirement of storage, the cup gains subjective concinnity. This "form follows function" feature alone would justify the innovation for many consumers.
- It is ergonomically more compatible than conventional handles. More of the handle's S-shaped surface conforms with the user's fingers and thus spreads the load over a larger area for greater comfort. The handle also provides a comfortable depression at the top for resting the thumb. Con-

Stack of Corelle cups

formity is technically a matter of objective concinnity, but realizing this after picking it up or seeing a picture of someone holding it increases subjective concinnity.

- It holds the fingers away from the bowl of the cup to avoid uncomfortable contact when it is hot.

- It requires less effort to hold than the "pinch grip" required by traditional cups. In fact, you can relax your hand almost entirely while holding a full cup of coffee, even with your thumb removed from the handle.

Aesthetically, the handle's empathic expression provides subjective concinnity in at least two symbolic ways:

- Its shape approximates the shape of a liquid stream under the influence of what is called the *Coanda effect,* the same physical phenomenon that causes a teapot to dribble after you've finished pouring. A surface tension causes the stream to cling momentarily to the mouth of the spout and curl back toward the vessel, as shown in the high-speed photograph here. Thus it is symbolically appropriate for a cup which, as the photo shows, tends to produce the same effect. While we might not be aware of the principle and our vision is not quick enough to capture the effect as clearly as a camera, we probably know, intuitively, that something like this happens. Thus it makes the similar shape of the handle seem appropriate and increases its subjective concinnity.
- Visual metaphors also reinforce the cup's subjective concinnity. This cup's *lip,* encircling its *mouth,* is quite literal; it has a fullness much like a per-

son's lip. Both metaphors seem natural, too, because cups are used in contact with the user's mouth and lips. The shape of the handle invites an easy mental jump to another mouth-related metaphor—it alludes, quite literally, to a tongue. Again, the allusion is quite literal. It has the same channel running its length, flanked by swelling along its edges.

GENERAL STRATEGIES

The designer has many ways of imbuing a design with subjective concinnity. Essentially anything in the product's form that brings to mind something that makes sense to the viewer in connection with the product will do. Listed in decreasing order of strategic importance, they include

- Daimons
- Stereotypes
- Ideals
- Agency
- Zeitgeists
- Nostalgia
- Clichés

Daimons, Stereotypes, and Ideals

Daimons, stereotypes, and ideals are the most crucial sources of subjective concinnity. Designers can effectively measure and evaluate all three with semantic differential surveys. They also can use these surveys as guides for optimizing daimons through refinements of design.

The design of this iron began with the accompanying empathic sketch. The designer wanted to express the most crucial source of subjective concinnity, the daimonic essence of the ideal iron, which happens to include a sense of smoothness and cleanliness.

When a new product's design brings to mind an appropriate daimon—when an iron looks like an iron and a computer looks like a computer—the product quickly makes sense and the affective-cognitive lag tends to be brief. A semantic differential survey confirmed that the smoothly sweeping line did this.

A product's semantic profile (its daimon) can be evaluated only in the context of what Osgood called *semantic space.* For convenience, Osgood usually depicted it as a three-dimensional space defined by three orthogonal axes: evaluation, potency, and activity. He chose evaluation because half of all adjectives are evaluative in nature. He chose the other two because most of the remaining half are about evenly divided between adjectives that connote potency (or power) and activity. Most adjectives imply all three, in one mixture or another, but many are more aligned with one of the axes than the

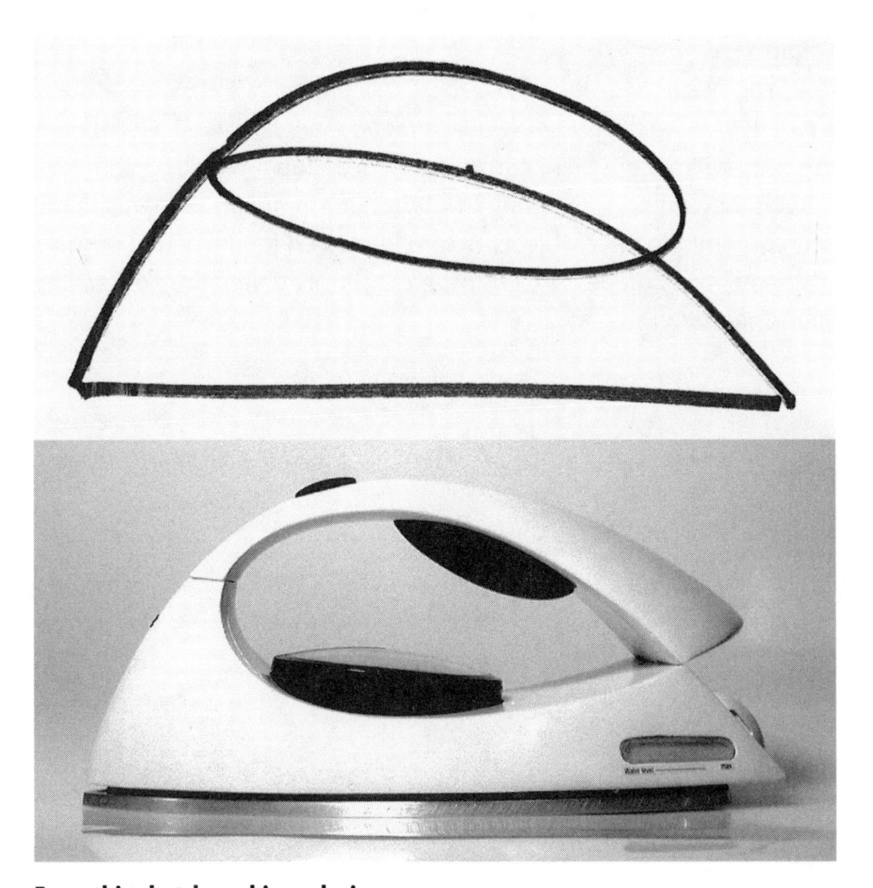

Empathic sketch and iron design

others. *Beautiful-ugly* lines up almost perfectly with the evaluative axis, as does good-bad. *Dynamic-static* lines up nearly parallel with the activity axis. *Strong-weak,* as you might guess, lines up with the potency axis. Every other scale corresponds to a vector passing through semantic space at some particular angle, uniquely determined by how closely its basic meaning corresponds to each of the dominant factors: evaluation, potency, and activity.

The three-dimensional average of the scores on all scales of a product's semantic profile defines a unique point in semantic space. This point distills the purest essence of the product's daimon—its meaning. The distance between it and any other object, similarly represented in the space, provides a quantitative measure of how they differ in meaning. The concept of a refrigerator resides at a less active and potent place in semantic space than the concept of an SUV.

Since we can determine the meaning of abstract concepts as well as real ones in the same way—like the *typical* cup and the *ideal* cup—we also can measure how much they differ in meaning from that of a *real* cup. Spheres in the illustration represent the three objects we are interested in: the actual product being considered (the Corelle cup, for instance, simply called the *design*), the *ideal,* and the *stereotype.*

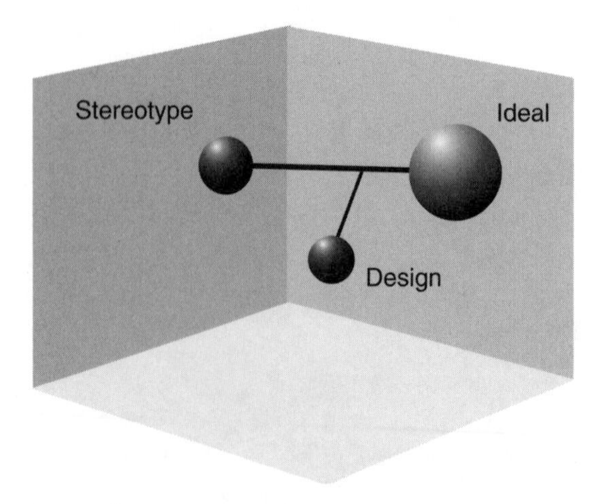

Semantic space

The sizes of the spheres correspond to the degree of consensus associated with each, as revealed by a survey. The sphere representing the design is relatively small because everyone evaluates one and the same real object; scores are not likely to vary much. The sphere representing the stereotype is larger because there generally will be less consensus about it; no two people have seen exactly the same mix of cups over their lifetimes. Furthermore, two subjects likely do not own the same kinds of cups, so the one they drink coffee from in the morning will be the one they see most frequently and saw as recently as the day they answered the survey.

The sphere representing the ideal is even larger and fuzzier. It is influenced not only by each subject's own cup but also by personal values, beliefs, attitudes, and concerns pertaining to the economy, the environment—anything and everything of personal importance, including the zeitgeist of the moment. To the extent that each person comes from a different family or a different culture, with different values, beliefs, and attitudes, the ideal is less distinct still.

Relationships among the three key concepts in semantic space define parameters of concern to the designer:

- The design's *novelty* is the distance between the stereotype and the design.
- The *design potential* inherent in the situation is the distance between the stereotype and the ideal, which defines the stereotype-ideal axis. It is a measure of the degree of novelty that survey respondents will tolerate. The stereotype and ideal will be close together if the group has conservative tastes and far apart if the group is avant garde. The further away the ideal is from the stereotype, the greater opportunity the designer has for creating a new and better design.
- The design's *subjective concinnity* corresponds to the shortest distance between the design and the stereotype-ideal axis. The closer the design lies to the axis, the more subjective concinnity it has. A design closer to the stereotype end of the axis should appeal more to conservative consumers; one closer to the ideal end should appeal to early adopters.

The designer optimizes the design's subjective concinnity—and thus its appeal—by changing its appearance in ways that move it closer to the stereotype-ideal axis. Long-term objectives are better served, however, by aiming for the ideal because that strategy leads to seminal forms. Companies that rely on focus groups and similar tools for gauging consumer preferences seldom end up with innovative product designs because such tools focus on the stereotype instead of the ideal. Answers to open-ended questions tend to steer the discussion toward what subjects know best—and what they agree most about—the stereotype. When the discussion turns to "blue sky" or "what if?" issues that promise to point to the ideal, the discussion gets ambiguous and fuzzy—naturally so because the ideal *is* a fuzzy concept. The facilitator frets about things getting out of control. In trying to regain control, he or she unwittingly steers the group back toward the stereotype again.

A comparison of the Corelle cup's semantic profile with the profile of an ideal cup shown earlier reveals, generally, that the designer did a good job with respect to subjective concinnity. The ugly-beautiful scale at the bottom, used for diagnosis, supports this assumption; the group judged the "beaver tail" cup to be quite beautiful. Incidentally, the ugly-beautiful scale was not used for the ideal cup survey because it would have amounted to a moot question; of course the ideal would be beautiful.

Generally, the diagnosis of positive valence is borne out by several results:

- The profile of the real cup closely shadows that of the ideal.
- While it falls slightly behind the ideal in a few cases regarding novelty—it seems less interesting, older, and more usual than desired—the shortcomings are so minor as to be insignificant and not worth the trouble to correct. Incidentally, I used three scales to test novelty (old-new, boring-interesting, and usual-unusual) for greater confidence in this important measure; three correlated opinions provide stronger evidence than one, which could have been a fluke.
- In other cases it exceeds expectations. It seems even lighter, simpler, and colder than expected of the ideal. Although it differs from the ideal on these scales, it differs in the direction of better than hoped for. The ideal would seem light and simple, for example; the actual cup seems even

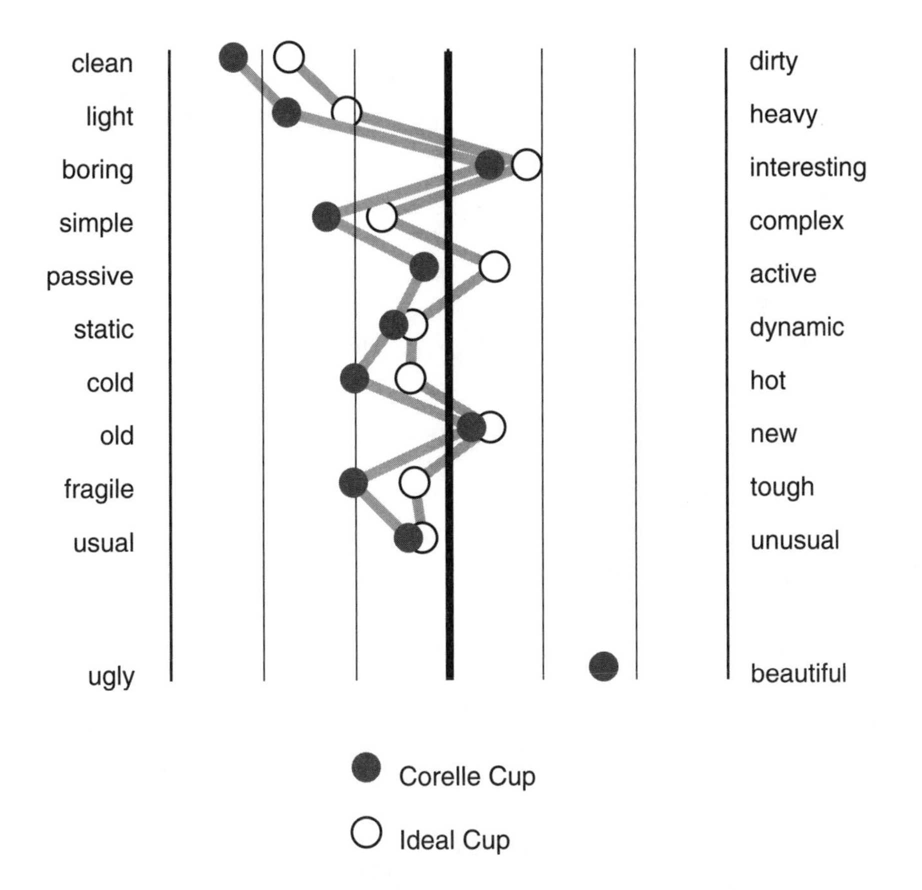

lighter and simpler. Such scores can be regarded as indications of a stronger expression of desirable character.

- The scales are ranked from top to bottom in order of their significance with respect to the ideal and their importance; the designer should heed those at top most while considering fixes. The most extreme score, and therefore the most significant quality, pertains to *cleanness.* Among the particular qualities used in this profile, it showed up as most important: Above all, a cup should seem *clean.* The Corelle cup actually exceeds the ideal in this regard, seeming even cleaner than expected.

So why isn't it still available? Although the rest of an expanded Corelle line is still in stores, the "beaver tail" cup was replaced by one with a more conventional handle. Apparently, the design's greatest shortcoming was an apparent fragility of the handle. Although it was made of a newly developed space-age combination of glass and ceramic that was virtually indestructible, people worried that the handle would snap off while they used it or that it would shatter if they dropped it.

Corning anticipated the problem and tried to blunt it with TV commercials that showed the cup dropping on a hard floor, from which it bounced, intact, like a ball and rang like a bell. The company also publicized a lifetime no-questions-asked guarantee; if one ever broke, a customer could return it for a free replacement. Apparently, the company could not sufficiently allay consumers' worries.

All of its subjectively concinnous features, whether revealed by comparison with the visual stereotype or the semantic profile of the ideal cup, refer to the core of its daimon—to its valued character *as a cup.* They do not depend on the context of temporary social or cultural norms. Its character will remain relevant for as long as people use cups the way they always have. The handle's form goes beyond symbolic reference to become truly an example of form following function. Its practical and ergonomic improvement makes it an epochal innovation; it provides a *new and better* way of holding a cup. It did not turn out to be an example of seminal form because no one else copied it—but it should have.

In practice, a designer using my methodology concentrates on making changes that move scores on individual scales of the semantic profile closer to the ideal—or *beyond it and even farther away from the stereotype.* It happens that moving one score often moves others in the same direction. This happens because of the way the scales are oriented in the graph. Although their left-right orientations were randomized during the survey, the graph shown here was adjusted so that the more active or potent term of each pair lies to the right. In other words, moving a score to the right is tantamount to moving the design farther out along the activity and potency axes of semantic space. Generally, the designer moves scores to the right by increasing information.

Adding information moves scores to the right on the following scales:

slow–fast

passive–active

calm–excitable

cold–hot

intentional–unintentional

simple–complex

straight–curved

blunt–sharp

boring–interesting

bland–pungent

plain–ornate

still–vibrant

ordered–chaotic

smooth–rough

relaxed–tense

gentle–violent

static–dynamic

rational–emotional

We now know that the designer can move scores on certain scales of a semantic profile to the right by adding information to a design in the form of either contrast or novelty. But how does he or she move them to the left? Reducing information would work. And sometimes this is the best or only way. However, removing information—and the aesthetic potential it contributes—should be done only as a last resort, especially if it means giving up hard-won novelty.

This is where the two kinds of concinnity interact to unify the subjective and objective notions of aesthetics: *Increasing objective concinnity, through application of the principles listed in Chapter 9, moves scores on certain scales to the left.* Thus objective concinnity directly affects subjective concinnity.

The leftward bias of the Corelle cup's scores indicated that it is already marked by substantial objective concinnity. So is the ideal, which suggests that we don't think of cups in general as powerful or active products. While such traits would spell disaster for a car—which is perceived as essentially active and powerful—they are apparently appropriate for a cup.

LESS INFORMATION
MORE OBJECTIVE CONCINNITY

Less Potent
Less Active

MORE INFORMATION
LESS OBJECTIVE CONCINNITY

More Potent
More Active

The permutations of the Corelle cup in this illustration were created by decreasing information in the form of contrast (left) and increasing it (right). Consequently, the one on the right seems more alive, potent, complex, and emotional. At the same time, the left cup has more objective concinnity, while the right one has less.

The designer has incentives for going beyond the ideal and farther away from the stereotype. Because the Corelle cup seems even cleaner than the ideal, the designer would have had to make it seem dirtier to make it come closer to the ideal; this just wouldn't make sense. It is better to have an exaggerated impression of cleanliness. It is also novel in this regard, and more exciting, by virtue of its greater distance from the stereotype.

Leaving that scale alone would be a good tactic for another reason: Moving a score in the right direction on one scale often moves other scores in the wrong direction. Let's suppose, for example, that the designer decided to improve the score on the boring-interesting scale by making the design just a little more interesting. He could do this by adding some contrast in the form of a more voluptuous profile curve. While the change would improve things on the boring-interesting scale, it also would move the score on the clean-dirty scale toward dirty. But not to worry—this time. In effect, the designer has money in the clean-dirty bank he can afford to spend on the boring-interesting improvement. It also would make the cup look more complex, but there's money in the simple-complex bank, too. Elsewhere, the extra contrast actually would compound the improvement; it would make the cup seem more active, for example, a desirable quality it is short on.

Agency and Zeitgeists

It would be great if product designers could accurately predict the next zeitgeist with enough lead time to catch it and ride it. If they could, all their products might become icons. Unfortunately, they cannot predict with much certainty what preoccupations will rise up to shape the next zeitgeist. Few people saw the Internet age coming, for example, until it was upon us.

One preoccupation seems everlasting, however, and predictable. We never lose a basic desire for what can best be called *agency,* from a Latin word meaning "to do." Think of agency as a sort of superfunctionalism. The maxim that "form follows function" is as venerable as any other guiding principle of design. A product not only should express what it is through its daimon, it also should express what it does.

However, agency has broader, more inclusive implications than functionalism. The form of a computer that looks like it computes well by virtue of its *precise* lines, among other things, has a form that follows its function. However, a "form that follows agency" also implies instrumentality or service in a more general sense. An *agent,* as we all know, is someone or something that helps a person achieve some desired end—with less physical or mental effort, in less time, with fewer resources, or at less expense. Agency is important because, like all living organisms, we spend physical and mental effort and material resources with reluctance. Whenever possible, we try to conserve our energy and other resources by relying on agents. When you want to buy or sell a house, you normally hire a real estate agent to handle all the nagging details. An employment agency takes on much of a job seeker's burden in finding and landing the right job. My literary agent made the job of finding a publisher for this book much easier than if I had attempted to do so myself.

Even before the first humanoid stuck a stick between two rocks to make a lever in order to move a third, our ancestors wanted to do more with less. And not with just human agents. I have already mentioned software agents that relieve computer users of mental tedium by taking care of routine tasks. Every tool, instrument, or product acts as an agent.

In general, a product's agency increases any time its utility, instrumentality, or benefit on behalf of its owner or user increases and/or any cost associated with it goes down. Cost includes more than purchase price. The physical or mental effort required of the user amounts to a cost. Thus, better ergonomics increases agency. The obvious presence of nonrecyclable materials represents a cost to the environmentally concerned consumer.

The "more for less" essence of agency embodies the scientific and mathematical notion of elegance and the economic notion of value. *Greater agency yields greater perceived elegance and value.* A Palm Pilot delivers more agency than an ordinary notebook and pen. A watch has more agency than a clock. A PT Cruiser has more agency than a New Beetle because it hauls more. A laptop computer has more agency than a desktop computer with the same power and display size because it does more for the user with less. Laptops not only get lighter and easier to carry; they also get thinner and the displays grow larger. Higher cost currently gives away some of this advantage. However, you also could say that a laptop commands a higher price because it provides, on balance, more agency. In any case, it seems to represent more value and elegance to the consumer. Laptops are thus more aesthetically exciting and appealing to consumers than desktop units, regardless of any other aspects of form.

Agency has such fundamental appeal that it recurs in the mythology of every culture. Aladdin rubs a lamp, and a powerful Genie appears to grant any wish. A flying carpet whisks its owner to a destination he has only to imagine. No muss, no fuss. Dorothy clicks the heels of her ruby slippers, and when she opens her eyes, she is back home in Kansas. Dick Tracy talks to his chief, no matter how far away he is, by talking to his watch. Commuters of the future will read or play games with fellow passengers while their cars are guided smoothly, safely, and effortlessly by magnets buried in the pavement and supercomputers under their hoods—or better yet, in flying cars that avoid that nasty congestion on the ground altogether.

Each of these fanciful products exemplifies ultimate agency. With the exception of Dick Tracy's wrist radio, products in the real world fall short of the ideal. However, each generation of products gets closer to the ultimate: Timepieces that began as large, cumbersome, and expensive machines eventually became small enough to wear on wrists and cheap enough for everyone to own. And they do not have to be wound anymore. Those which receive time signals via radio waves do not even have to be set. Cars do not have to be shifted anymore, and their windows do not have to be cranked. And we do not have to get off the couch to change channels on the TV anymore.

The history of any product class can be traced as a history of increasing agency. The future of any product class can be predicted as a continuation of the trend toward ever-increasing agency. Even before someone stuck a stick between rocks to made a lever, we have always wanted to do more with less. The general leverage of agency—reducing effort and costs while increasing effect and benefits—has been the earmark of what we call *progress.* In this sense, every age has been, at root, an *age of agency,* regardless of what else people happened to be preoccupied with and what we eventually named it: stone age, bronze age, age of steam, age of electricity, atomic age, or information age. Thus, agency is a good bet for improving the appeal of any product. Until we can achieve our grandest fantasy with the merest of wishes, we will never have enough agency. Regardless of what we call the next age, therefore, we can confidently predict that consumers will expect laptop computers that get smaller, lighter, faster, and cheaper. And they will expect their batteries to last longer.

Smaller size looms large in products of the information age, but larger size also can increase agency. A longer lever enables a person to move a heavier stone with less effort. Likewise, a larger, more powerful SUV will be perceived by SUV enthusiasts to have more agency and thus more value, especially if it is no more difficult to drive and no more expensive.

Invisibility or transparency is an important feature of agency. Ideally, you are never aware of an agent at work—only of its results. The people who wait on you at the best restaurants wait motionless in the background, in their black, relatively unnoticeable uniforms. They move only to make things appear on your table, or disappear, at just the moment you wish they would. Their motions are unobtrusive as they reach deftly around and between diners and table pieces, trying to be as unnoticeable as possible. When dinner is over, the slightest gesture summons the check. When large staffs of servants were customary in homes of the rich, the servants of the grandest houses not only lived in separate quarters, out of sight, but they also moved among floors and rooms through separate hallways and stairways, hidden from the family's view behind the walls.

Dick Tracy's two-way wrist radio enabled him to get in touch with headquarters without spending the time and effort looking for a phone booth. Today's cell phones do the same for us, and much more. Some even fit the wrist. Will we wear future phones like today's virtually invisible hearing aids, and will we dial them via bone conduction just by speaking the phone number? Or more likely, simply by saying, "Call Betsy"? Such a product is so predictable, given our insatiable appetites for agency, that I have no doubt it is on the drawing boards now—if not in prototype form. Will future phones be implanted in our heads and surgically connected to our brains so that I can merely think "Call Betsy"? It is probably inevitable. Not only does it promise more agency, movies have already prophesied it.

As I have already suggested, we are such creatures of our times that designers inevitably express the spirit of their time, the zeitgeist, in their work. It would be difficult, in fact, not to reflect it. In the unlikely event that a designer could ignore the zeitgeist, the resulting product would look "out of place" or "ahead of its time." If a product's "form follows its function," it evokes its beckoning ideal. It conforms as well to the universal ideal of agency when it does more while requiring less effort of the user and appears in a much slimmer, lighter, and smaller package than usual.

Clichés and Nostalgia

Clichés are closely related to stereotypes but less restrictive because they allow the designer to borrow from stereotypes of other product categories or even from nonproducts. Cars have clichés from aircraft and computers. They represent the easiest, most obvious way—the no-brainer way—for designers to ensure that their designs have subjective concinnity. Like the mystery writer who begins with "It was a dark and stormy night . . . ," the designer does not have to think long or hard in order to use a cliché. In fact, clichés just appear, as if by magic, precisely when a designer is on autopilot and gives little or no thought to matters.

Many clichés arrive with seminal forms that change stereotypes and the ground rules for designing similar products that follow. The decision to use transparent and bright, anything-but-gray colors in the design of a high-tech product was cause for weighty deliberation—and probably debate—in the studios and offices at Apple. Such novel departures carry considerable risks. Once the iMac succeeded, however, this simple, inexpensive innovation became an asset that virtually any manufacturer could capitalize on. Others might be accused of following rather than leading, but they could take solace in being "with it."

Predictably, peripherals designed for use with the iMac soon sported transparent colors to match the iMac's. However, many other colorful, transparent products quickly appeared as well, ranging from cellular phones to office chairs and virtually everything in between. Once this happened, transparent color had become a cliché—just another easy way to make a product seem stylish, yet normal.

As easy and safe as clichés are to employ, however, they earn the designer and manufacturer no bankable credit; the only one to benefit from transparent colors in the long run will be Apple, for whom it was not a cliché but a seminal innovation. Apple's products benefited immediately from the freshness, precision, and richness of crystal-like transparency. It continues to benefit from its bolstered image as a leader. Furthermore, every subsequent transparent emulation reinforces Apple's brand equity by reminding all who see it of who started it—even if only subliminally. As the final payoff, only Apple's products, among all other transparent products, will be recalled in design history for their innovation and mounted in future museum exhibits. The emulators will be remembered chiefly as opportunists and Johnny-come-latelies. While clichés make good short-term investments, they make poor strategic investments.

Incidentally, the iMac's image benefits, too, in the company of color-coordinated peripherals. Each one increases its own objective concinnity by repeating the iMac's color and transparency. However, they all increase the iMac's objective concinnity and sense of place at the same time. They establish a sense of unity in the whole ensemble.

Volkswagen's New Beetle, DaimlerChrysler's PT Cruiser, and Ford's latest Thunderbird depend on concinnity derived from nostalgia that reminds viewers fondly of revered cars of the past. Many watches rely on past models for inspiration, as does every piece of period furniture.

Timeless Design

Despite its name, museums do not display Movado's "Museum Watch" because it is derivative of Horwitt's creation, the real museum watch. However, it is the only way to personally enjoy what Horwitt did. The finned cars of the 1950s make sense only in the historical context of their exuberant and paranoid decade. They are interesting to contemplate historically and archaeologically for a better understanding of the period. And they might show up in special art museum exhibitions for this reason. However, they likely will not end up in permanent collections (except in car museums). Likewise, even though the original VW Beetle that inspired it is already a classic, the New Beetle probably will not become one.

Winning Design

11

WE HAVE ARRIVED at a point where I can now summarize a recipe for serving up winning product designs that appeal aesthetically to the broadest number of people—or at least to consumers the manufacturer intends to court. Following the recipe faithfully yields designs of relatively timeless appeal that outlive short-term fashions and fads. Following it consistently should increase the value of any brands associated with the resulting products.

The winningest designs aren't merely new; they embody epochal innovations and seminal forms that compel competitors to follow suit:

- *Epochal innovation,* a concept due to Harvard's William Abernathy, establishes a new and better way of doing something with, I would add, functional or ergonomic implications.
- *Seminal form* establishes a new visual trend, norm, or class.

Epochal innovation and seminal form can be as shocking as any other form of novelty, but they are unique in containing the seeds of their own redemption. At the same time they depart from one or more stereotypes, they always depart in the direction of one or more ideals. They constitute the most powerful of all aesthetic resources, *concinnous innovation.* In addition to their ability to arouse the interests and passions of the viewer, these self-contained packages of aesthetic dynamite also carry with them a relatively stable form of subjective concinnity that makes sense of the innovation, either actually or symbolically.

As components of a pervasive mass medium, they transcend entertainment to edify their audiences; they don't merely please by reinforcing existing, often inappropriate stereotypes. Instead, they challenge audiences and inform them literally by changing minds and opening eyes to new and more appropriate ways of seeing the world—as the Horwitt watch, the Palm Pilot, and the 1986 Ford Taurus did. Such designs establish or maintain reputations for leadership that tend to insulate their sponsors from the risks of even more daring innovations in the future.

JUST TWO AESTHETIC FACTORS

In the most general sense, only two factors determine an object's aesthetic nature: *information,* which accounts for its arousal potential, and *concinnity,* which accounts for its valence. The inherent information in an object's form determines how exciting and interesting it is. The more information it embodies, the greater is its ability to capture attention and stir feelings, passions, and urges.

For the object to be attractive, however, the information also must make sense—and rather quickly. This depends on the degree of concinnity simultaneously present in the object's form. Any feature or arrangement of features that enables the nervous system to process and make sense of the information is concinnous by definition.

JUST FOUR INGREDIENTS

In particular, the recipe for winning design calls for the designer to blend just four ingredients: *contrast* (objective information), *novelty* (subjective information), *objective concinnity,* and *subjective concinnity.*

- *Contrast* arises from obviously different attributes (different colors, textures, line curvatures, surface change, etc.) that we can measure objectively with instruments such as light meters and rulers.
- *Novelty* arises from perceived differences between a real product and its stereotype, an imaginary (subjective) mental model that the product automatically brings to mind. The stereotype represents what the viewer *expected* the product to resemble. Stereotypes are not directly accessible and measurable in the same ways real products are; they can be measured only with psychological instruments, such as semantic differential surveys that tap the viewer's mindset.

- *Objective concinnity* arises from similarities among shapes, colors, dimensions, textures, and other visual attributes. Like contrast, it can be measured objectively with instruments such as light meters and rulers.
- *Subjective concinnity* arises from similarities among a product, its stereotype, and its ideal—another mental model that automatically springs to mind. Unlike the stereotype, which corresponds to what the viewer most *expected,* the ideal corresponds to what the viewer implicitly *hoped* the product would resemble.

These four attributes represent the only aesthetic tools available to the designer for adjusting a design's arousal potential and valence. Fortunately, they are also the only ones needed.

A MATTER OF BALANCE

We can better understand how the four attributes interact by visualizing them in the context of hypothetical scales for measuring arousal and valence.

The arousal scale actually weighs information. Increasing a design's contrast or novelty increases the information-processing load on the viewer's nervous system, which is reflected in an increase of the psychological and physiologic arousal we normally call *excitement.* It gets and holds attention better. It becomes more interesting and *moving.* By the same token, reducing contrast or novelty makes a design less noticeable and less exciting.

Arousal scale

Three valences are possible, depending on the relative balance of information and concinnity. All the terms we use to describe positive valance have positive connotations: attractive, beautiful, delightful, interesting, likable, charming, fetching, fascinating, exciting, enchanting, engaging, entrancing, captivating, intriguing, and provocative. As a figurative rule, the positive valence of a pleasant and attractive product requires a balance of information and concinnity. Any increase of information, in the form of either contrast or novelty, must be counterbalanced by a corresponding increase of either objective or subjective concinnity in order to maintain positive valence.

All the terms we associate with the unpleasantness of negative valence have negative connotations, of course: unattractive, ugly, repulsive, repellent, repugnant, revolting, abhorrent, offensive, disgusting, obnoxious, hideous, loathsome, horrid, and awful. Lacking sufficient concinnity, an ugly design doesn't make sense; it is literally *meaningless.* A design that depends primarily on novelty for its affect, with little concinnity to back it up, can be thought of "novelty without redeeming value."

A design tends toward neutral valence when concinnity outweighs information, with the result that it doesn't necessarily attract or repel viewers. They

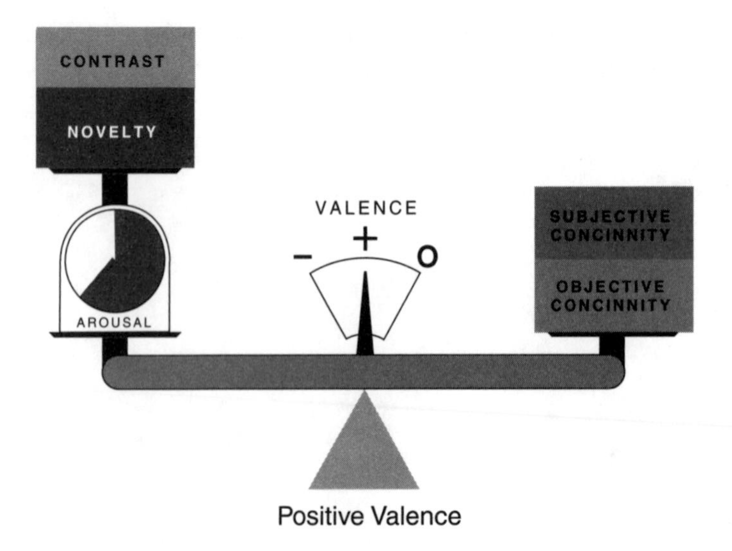

Positive valence

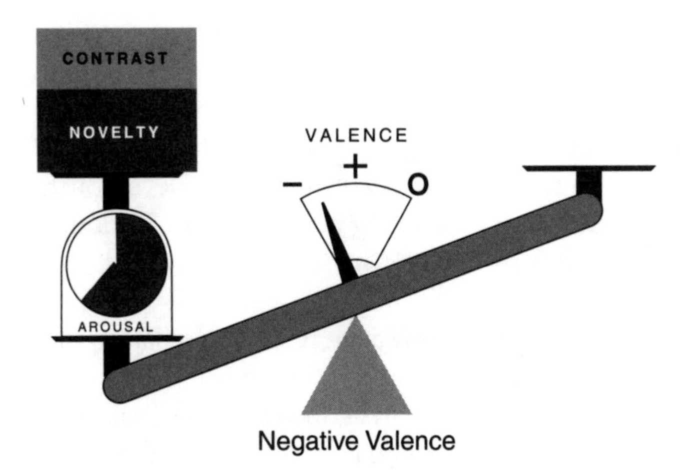

Negative valence

Negative Valence

are more likely to feel indifferent toward it—or to not notice it at all. Products with neutral valence seem normal, typical, and predictable—like stereotypes.

Products characterized by neutral valence occupy an evaluative middle ground where descriptions have both positive and negative connotations. Positive terms include simple, compatible, friendly, clean, unobtrusive, harmonious, tasteful, nice, and coherent. Negative terms include dull, uninteresting, ordinary, unexciting, plain, lackluster, common, and humdrum.

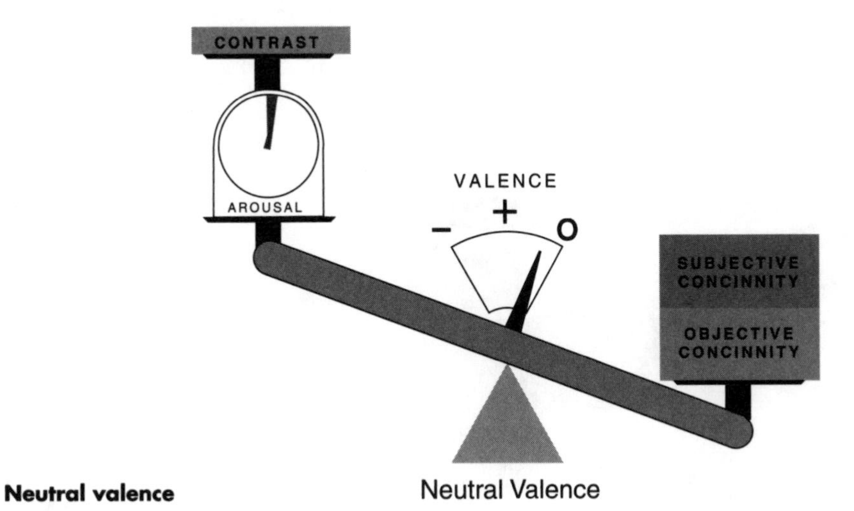

Neutral valence

Neutral Valence

Carried to extremes of the most subtle contrast and least novelty, a product can have so little arousal potential that it might blend unnoticeably with its surroundings. This does not bode well for a product that must elbow its way among competing products on a store shelf to dominating the attention of consumers. Nevertheless, Henry Dreyfuss regarded unobtrusive products, which he called *classic* designs, as the crowning achievements of industrial design: "A telephone should never be noticed," he said, "until it rings."

While neutral valence has less ability to actively attract consumers than positive valence, it is certainly preferable to negative valence, which commands attention but drives consumers away. In the final analysis, though, neutral valence is more important to users than consumers because it approaches the ergonomic ideal of *anesthetic* design. The normal and predictable look of neutral valence is equivalent to the ergonomic concept of compatibility. It has little potential for distracting or confusing users with irrelevant, noisy information. Neutral valence is especially important in the design of products such as computers that exist for the primary purpose of conveying essential information. It is most valu-

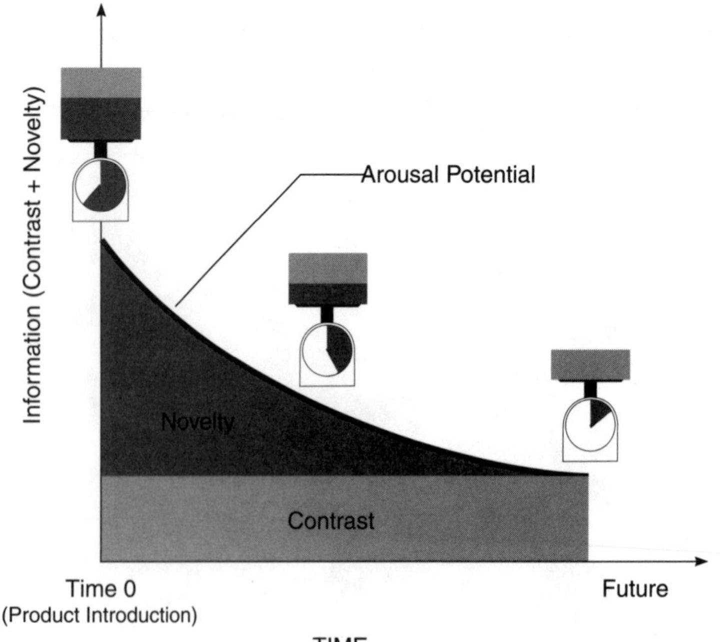

able when applied to the blank slate on which essential information is written. While the back side of a computer's monitor might have consumer-pleasing positive valence—by virtue of ample amounts of discretionary contrast and novelty—its screen should present the plain look of neutral valence in order not to interfere with more important essential and collateral information.

STRATEGIC CONSIDERATIONS

Technically, a designer can achieve positive valence with any mix of contrast and novelty on the information side of the scale (although novelty is far more potent, due to its instinctive association with environmental threats) and any mix of objective and subjective concinnity on the other side. However, some mixes are more effective strategically because they produce more predictable or stable results.

Transient Novelty

All designs must embody some contrast. While a designer might effectively eliminate novelty from a design for all practical reasons, no object could be perceived without at least a modicum of contrast. Otherwise, we could not perceive it at all. Contrast never changes, either. The contrast of two colors will be the same tomorrow or a hundred years from now as it is today (disregarding the natural degradation due to age, of course). Thus its effects on valence are utterly timeless, stable, and predictable.

Novelty, however, long ago proved itself a more potent consumer stimulant than contrast. Thus virtually every product is designed to differ somehow from its predecessors and established norms when it first appears. The problem with novelty over the long run, of course, is that nothing remains new forever. The novelty of an unusual product begins to dissipate, taking arousal potential with it, from the first moment a consumer lays eyes on it.

This inevitable tendency to lose its punch explains, in part, how and why some preferences change over time. Assuming for the moment that an unusual product's concinnity remains constant (which, in fact, it seldom does), the valance scale begins to tip in the direction of neutral valance as the load of information lightens and the arousal scale unwinds. The product seems progressively more ordinary and less exciting in the process. When novelty is the primary source of arousal, a product that might have a dramatically new look at the outset might end up a crashing bore.

But familiarity doesn't always breed contempt. Something even more dramatic can happen when a product starts out with negative valence due to a surplus of novelty. Assuming, again, that concinnity does not change, the dissipation of novelty causes the scale to rotate toward a state of *positive* valence. Thus a product that seemed ugly at first actually can come to seem beautiful, given enough time. We can like even the most abhorrent things if exposed to them long enough.

I can demonstrate the effect with an "ugly" design created to illustrate a magazine article about automotive aerodynamics I wrote in 1980. The hypothetical car on the right incorporated minor modifications of the 1980 Chevrolet Citation designed to increase its aerodynamic efficiency enough to boost fuel economy some 15 percent: relatively circular sweeps of the windshield and nose, a sloping nose, and skirts covering the rear wheels, among other details. While it looks rather ordinary and uninteresting today, bear in mind that fashionable cars of the period were boxy and crisp-edged, not roundish and soft-

1980 Chevrolet Citation and Aero Citation

edged. My illustration appeared a full 5 years before the seminal 1986 Ford Taurus shocked Americans with the "aero look" and revolutionized car design. Not even the similar 1983 Audi 5000 had appeared yet. So my design seemed so strange at the time that my editor was reluctant to publish it: "That's the ugliest car I ever saw," he said. In an effort to appease him, I increased its subjective concinnity by adding whitewall tires, a cliché that was still popular.

I thought it was beautiful from the beginning, of course, because all its novel features made sense to me; for me, it had plenty of concinnity. My editor probably came to like it more over the years as its novelty wore off, perhaps even to the point of thinking it was beautiful, especially after he became accustomed to the Taurus. If a product's novelty happens to be epochal or seminal in nature, as in the case of the original Taurus, the transition toward normalcy can accelerate as it also gains subjective concinnity. This gain occurs, in part, as the competition is compelled to emulate the new design. As the field becomes populated with lookalikes, the once unusual product seems typical even more quickly. In effect, the automotive stereotype gravitated toward the Taurus. That is, it redefined its whole category of products. My design assumed its rather ordinary look in the process, along with the Taurus. Except, perhaps, for the skirted rear wheels; they might still be unusual enough to disturb some viewers.

It is one thing to design a product to look normal at the outset by emulating some existing stereotype. It is quite another thing to have it end up looking normal by forcing the stereotype to conform more closely to it. Stereotype-altering design (epochal innovation and seminal form) defines design leadership—which can mitigate any loss of excitement due to fading novelty. A more cognitive kind of excitement can compensate the loss as a product's owner realizes that he or she owns a trendsetter.

Deep Concinnity

A product's valence always changes in predictable fashion to the extent that it depends on transient novelty. Valence usually changes as well to the extent that it hinges on subjective concinnity—but often in unpredictable ways. In most cases, the degree of a product's subjective concinnity remains in doubt because many of the subjective forces that determine subjective concinnity—values, beliefs, concerns, etc.—remain in flux. And of course, there is always the possibility that a competitor will alter standards overnight by introducing its own epochal innovation or seminal form.

Because expectations, attitudes, values, and other aspects of a person's mindset vary to some extent from those of other people, what is novel to one person will be more or less novel to others. Novelty always diminishes over time, as the product's look becomes more familiar.

Stereotypes and ideals differ from one person to the next. And they change over time. So a design's subjective concinnity will differ among different people, especially those of different cultures. It can increase or decrease over time as circumstances, concerns, values, and beliefs change.

Clichés, as sources of subjective concinnity, are especially capricious. Transparent colors, popularized by the Apple iMac, found widespread application in a remarkable range of products, including such unlikely products as office chairs. But they and their potential for lending subjective concinnity to a product lasted only about 2 years. Today, many of those products are likely to evoke only smiles, if not chuckles.

As for the other sources of subjective concinnity discussed in Chapter 10, the accompanying graph ranks them according to their relative long-term reliability and thus their general suitability. Clichés, the least reliable source of concinnity, occupy the shallowest, most volatile level of the hierarchy. At the other extreme, objective concinnity occupies the deepest level. Being more constant and reliable than all sources of subjective concinnity, it amounts to the virtual bedrock of all sources of concinnity.

Like its complement, contrast, it never changes. The objective concinnity inherent in parallel or vertical lines or symmetry will remain the same forever. Our resonance with objective concinnity is so instinctive in nature that it might well stem from hardwired circuits in our nervous systems. Future nervous systems will respond to it in the same ways ours do. This means that ob-

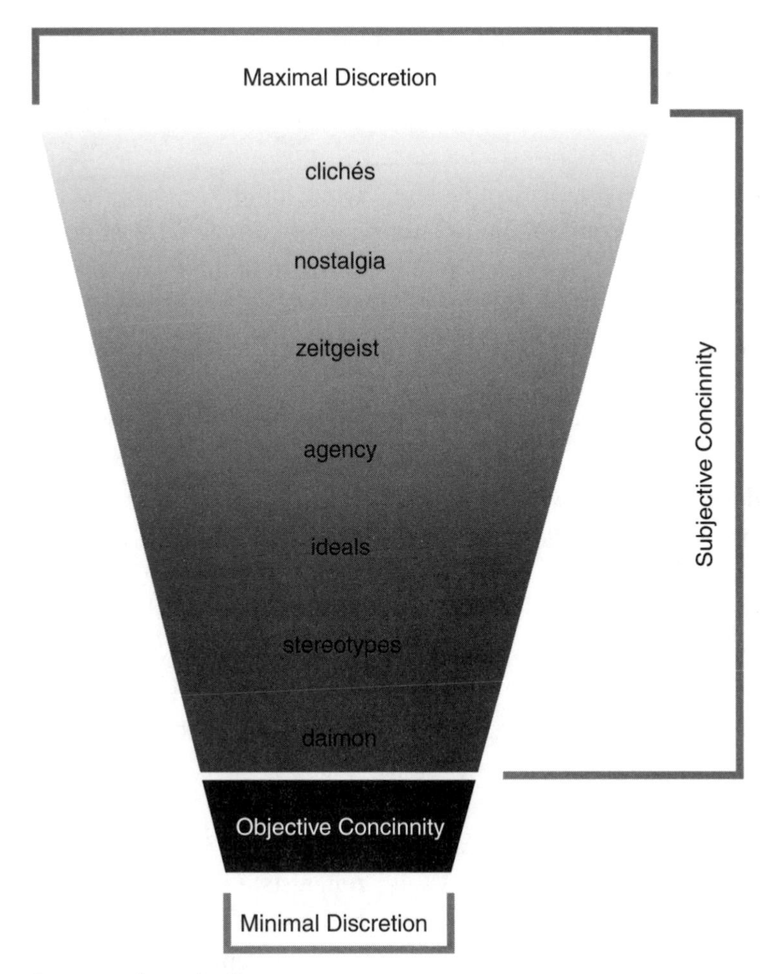

Sources of concinnity

jective concinnity, balanced by equally steadfast contrast, yields the most stable valence possible. Consequently, the most timeless designs—the museum pieces, the classics, the products that appeal to our sensitivities virtually forever—are characteristically marked largely by contrast and ample amounts of objective concinnity.

It is impossible to ignore subjective concinnity altogether. As I discussed in Chapter 9, manipulation of objective concinnity always affects subjective concinnity as well. Increasing objective concinnity, for instance, makes a design seem more rational and less emotional. This might increase the subjective concinnity of a computer, which we expect to seem rational, but decrease it in the case of a sports car, which we expect to seem more spirited.

The tapered overlay of the graph signifies the relative latitude a designer has in tapping various sources of concinnity and the approximate ease of doing so. It is widest at the top and signifies that clichés represent the most convenient source of concinnity. Indeed, they are hard to avoid. The designer has so much latitude at this shallow level, in fact, that he or she has the option of not using clichés at all. The extreme latitude also gives him or her license to draw on clichés from anywhere, even outside a product's own category—as in the case of transparent colors spawned by the iMac's. The designer has the least latitude at the deepest level. The means for achieving objective concinnity are very tightly defined. And a design with little objective concinnity almost surely will fail aesthetically. I have seldom seen a product that could not be improved with more objective concinnity. But, then, I am an industrial designer. Like many of my colleagues, I usually can't get enough. Nor can museum curators. Designers who want to create revered classics that end up in museum collections and win design awards also risk boring hordes of consumers who prefer more emotional, less refined designs.

Daimons are more reliable than clichés, and more crucial. A watch should seem precise. A car should seem agile. While zeitgeists are less fickle than clichés, they are nevertheless more changeable and less reliable than a good daimon or objective concinnity. Soft, "aero" shapes dominated automotive form a few years ago. Today, with the rampant popularity of SUVs, crisp, edgier themes are in vogue. Nevertheless, sleek sedans and rugged SUVs alike respect a more fundamental daimonic image: They all seem fast, powerful, safe, and agile if we like them.

Products of epochal innovation and seminal form—complemented with concinnity from as deep in the concinnity well as possible—yield the greatest long-term benefits. Manufacturers that follow design strategies based on concinnous innovation become trendsetters and leaders by forcing competitors to emulate them. Corporate prestige increases. So do brand and stock equities. Their products rise to foremost positions in the public mind, set standards, and establish new classes. From such heights, a manufacturer can try even bolder innovations with increasing confidence and impunity. The manufacturer

- Can introduce more daring functional and aesthetic innovations with less fear of recrimination from consumers and users
- Can enjoy earlier sales success in the life cycle of even the most radical product
- Can enjoy greater immunity from occasional missteps

BIBLIOGRAPHY

Arnheim, Rudolf. *Art and Visual Perception: A Psychology of the Creative Eye.* Berkeley: University of California Press, 1974.

Baars, Bernard J. *In the Theater of Consciousness: The Workspace of the Mind.* Oxford, England: Oxford University Press, 1997.

Banham, Reyner. *Theory and Design in the First Machine Age,* 2d Ed. New York: Praeger, 1967.

Bayer, Herbert, et al., (eds.). *Bauhaus.* New York: Museum of Modern Art, 1938.

Bel Geddes, Norman. *Horizons.* New York: Dover Publications, 1977.

Bennis, W. G., and B. Nanus. *Leaders: The Strategies for Taking Charge.* New York: Harper & Row, 1985.

Berlyne, D. E. *Aesthetic and Psychobiology.* New York: Appleton-Century-Crofts, 1971.

Berlyne, D. E. *Conflict, Arousal, and Curiosity.* New York: McGraw-Hill, 1960.

Berlyne, D. E. (ed.). *Studies in the New Experimental Aesthetics.* Washington: Hemisphere, 1974.

Bernstein, Leonard. *The Unanswered Question: Six Talks at Harvard.* Cambridge, MA: Harvard University Press, 1976.

Birkhoff, G. D. *Aesthetic Measure.* Cambridge, MA: Harvard University Press, 1933.

Blackmore, Susan. *The Meme Machine.* Oxford, England: Oxford University Press, 1999.

Boone, Louis E. *Classics in Consumer Behavior.* Tulsa, OK: The Petroleum Publishing Company, 1977.

Boulding, Kenneth E. *The Image: Knowledge in Life and Society.* Ann Arbor: University of Michigan Press, 1969.

Bragdon, Claude. *The Beautiful Necessity: Architecture as "Frozen Music."* New York: Quest, 1978.

Bruner, Jerome S. *On Knowing: Essays for the Left Hand.* New York: Atheneum, 1969.

Bush, Donald J. *The Streamlined Decade.* New York: George Braziller, 1975.

Chen, Shenchang. "A Computer-Assisted Approach to Shape Averaging and its Applications to Industrial Design." M.A. Thesis, Ohio State University, 1986.

Cheney, Sheldon, and Martha Cheney. *Art and the Machine: An Account of Industrial Design in 20th-Century America.* New York: McGraw-Hill, 1936.

Coates, Del. "Compatible Environments: Suggested Areas of Research." In *Proceedings, 16th Annual Conference of the Human Factors Society,* Los Angeles, October 17–19, 1972, pp. 1–7.

Coates, Del. "In Search of the Sense of Beauty." *Design Journal,* Fall 1972, pp. 8–13.

Coates, Del. "Understanding Aesthetics: From Old Shoes to a Teacup." *Industrial Design,* September-October 1979, pp. 32–35.

Coates, Del. "Measuring Product Semantics with a Computer." *Innovation,* Fall 1988, pp. 7–10.

Coates, Del. "CAID Currents: Information and Aesthetics." *Innovation,* Spring 1992, pp. 10–13.

Coates, Del. "Analyzing and Optimizing Discretionary In*formation.*" *Design Management Journal.* Spring 1997, pp. 58–65.

Coates, Del. "Aesthetics." In *Proceedings of the Human Factors Engineering Summer Conference,* The University of Michigan, Ann Arbor, 1970–2001.

Collingwood, R. G. *The Principles of Art.* Oxford, England: Oxford University Press, 1938.

Coyne, R. D., et al. *Knowledge-Based Design Systems.* Reading, MA: Addison-Wesley, 1990.

Damasio, Antonio. *Descartes' Error: Emotion, Reason, and the Human Brain.* New York: G. P. Putnam's Sons, 1994.

Damasio, Antonio. *The Feeling of What Happens: Body and Emotion in the Making of Consciousness.* New York: Harcourt Brace, 1995.

Dennett, Daniel C. *Kinds of Minds: Toward an Understanding of Consciousness.* New York: Basic Books, 1996.

De Sousa, Ronald. *The Rationality of Emotion.* Cambridge, MA: MIT Press, 1990.

Devlin, Keith. *Goodbye Descartes: The End of Logic and the Search for a New Cosmology of the Mind.* New York: Wiley, 1997.

Dissanayake, Ellen. *Homo Aestheticus: Where Art Comes From and Why.* Seattle: University of Washington Press, 1995.

Dodwell, Peter C. *Visual Pattern Recognition.* New York: Holt, Rinehart and Winston, 1970.

Dohrn-van Rossum, Gerhard (translated by Thomas Dunlap). *History of the Hour: Clocks and Modern Temporal Orders.* Chicago: University of Chicago Press, 1992.

Dougherty, D. "Understanding New Markets for New Products." *Strategic Management Journal,* Vol. 11, 1990, pp. 59–78.

Dougherty, D. A. "Practice-Centered Model of Organizational Renewal Through Product Innovation." *Strategic Management Journal,* Vol. 13, 1992, pp. 77–92.

Eckblad, G. *Scheme Theory: A Conceptual Framework for Cognitive-Motivational Processes.* London: Academic Press, 1981.

Engel, James F., et al. *Consumer Behavior.* Chicago: Dryden Press, 1990.

Ewen, Stuart. *All-Consuming Images: The Politics of Style in Contemporary Culture.* New York: Basic Books, 1988.

Fischler, Martin A., and Oscar Firschein. *Intelligence: The Eye, the Brain, and the Computer.* Reading, MA: Addison-Wesley, 1987.

Fiske, John. *Introduction to Communication Studies,* 2d Ed. London: Routledge, 1990.

Fitts, P. M., and C. M. Seeger. "S-R Compatibility: Spatial Characteristics of Stimulus and Response Codes." *Journal of Experimental Psychology,* Vol. 46, 1953, pp. 199–210.

Flavell, John H. *The Developmental Psychology of Jean Piaget.* New York: Van Nostrand, 1963.

Foucault, Michel. *The Archaeology of Knowledge and the Discourse on Language.* New York: Pantheon, 1972.

Frank, Thomas. *The Conquest of Cool: Business Culture, Counterculture, and the Rise of Hip Consumerism.* Chicago: University of Chicago Press, 1997.

Fraser, J. T. *Of Time, Passion, and Knowledge: Reflections on the Strategy of Existence.* New York: George Braziller, 1975.

Frondizi, Risieri. *What Is Value? An Introduction to Axiology,* 2d Ed. LaSalle, IL: Open Court Publishing Company, 1971.

Fry, Roger. *Vision and Design.* Cleveland: Meridian Books, 1920.

Gardner, Howard. *Frames of Mind: The Theory of Multiple Intelligences.* New York: Basic Books, 1983.

Gardner, Howard. *The Mind's New Science: A History of the Cognitive Revolution.* New York: Basic Books, 1985.

Gershon, Michael D. *The Second Brain: The Scientific Basis of Gut Instinct and a Groundbreaking New Understanding of Nervous Disorders of the Stomach and Intestine.* New York: Harper Collins, 1998.

Getzels, Jacob W., and Mihaly Csikszentmihalyi. *The Creative Vision: A Longitudinal Study of Problem Finding in Art.* New York: Wiley, 1976.

Ghyka, Matila. *The Geometry of Art and Life.* New York: Dover, 1977.

Gibson, James J. *The Senses Considered as Perceptual Systems.* Boston: Houghton Mifflin, 1966.

Gilbert, Katharine Everett, and Helmut Kuhn. *A History of Esthetics.* New York: Dover, 1939.

Ginsburg, Douglas H., and William J. Abernathy (eds.). *Government, Technology, and the Future of the Automobile.* New York: McGraw-Hill, 1980.

Goleman, Daniel. *Emotional Intelligence: Why It Can Matter More than IQ.* New York: Bantam Books, 1995.

Gorb, P. "Design as a Corporate Weapon." In P. Gorb (ed.), *Design Management.* New York: Van Nostrand Reinhold, 1990, pp. 67–80.

Gorb, P. "Introduction: What Is Design Management?" In P. Gorb (ed.), *Design Management.* New York: Van Nostrand Reinhold, 1990, pp. 1–9.

Gregory, R. L. *Eye and Brain: The Psychology of Seeing.* New York: McGraw-Hill, 1966.

Hargittai, István, and Magdolna Hargittai. *Symmetry: A Unifying Concept.* Bolinas, CA: Shelter Publications, 1994.

Haviland, Jeanette, and Mary Lelwica. "The Induced Affect Response: 10-Week-Old Infants' Responses to Three Emotional Expressions." *Developmental Psychology,* Vol. 23, 1987, pp. 97–104.

Hebb, D. O. "Drives and the CNS (Conceptual Nervous System)." *Psychological Review,* Vol. 62, 1955, pp. 243–254.

Hebb, D. O. *Organization of Behavior.* New York: Wiley, 1949.

Helson, Harry. *Adaptation-Level Theory: An Experimental and Systematic Approach to Behavior.* New York: Harper & Row, 1964.

Hochman, Elaine S. *Bauhaus: Crucible of Modernism.* New York: Fromm International, 1997.

Hoffman, Donald D. *Visual Intelligence: How We Create What We See.* New York: Norton, 1998.

Hofstadter, Albert, and Richard Kuhns (eds.). *Philosophies of Art and Beauty: Selected Readings in Aesthetics from Plato to Heidegger.* New York: Modern Library, 1964.

Huizinga, Johan. *Homo Ludens: A Study of the Play Element in Culture.* Boston: Beacon, 1955.

Huntley, H. E. *The Divine Proportion: A Study in Mathematical Beauty.* New York: Dover, 1970.

Izard, Carroll, et al. *Emotions, Cognition, and Behavior.* Cambridge, England: Cambridge University Press, 1984.

James, William (edited by Gordon Allport). *Psychology: The Briefer Course.* New York: Harper & Row, 1961.

Johnston, Victor S. *Why We Feel: The Science of Human Emotions.* Reading, MA: Perseus Books, 1999.

Jones, Russell A. *Self-Fulfilling Prophecies: Social, Psychological, and Physiological Effects of Expectancies.* Hillsdale, NJ: Lawrence Erlbaum Associates, 1977.

Kanter, R. M. *The Change Masters.* New York: Simon and Schuster, 1983.

Kantowitz, B. H., and R. D. Sorkin. *Human Factors: Understanding People-System Relationships.* New York: Wiley, 1983.

Kotler, Philip. *Marketing Management.* Englewood Cliffs, NJ: Prentice-Hall, 1991.

Kreitler, Hans, and Shulamith Kreitler. *Psychology of the Arts.* Durham, NC: Duke University Press, 1972.

Lakoff, George. *Women, Fire, and Dangerous Things: What Categories Reveal about the Mind.* Chicago: University of Chicago Press, 1987.

Lakoff, George, and Mark Johnson. *Metaphors We Live By.* Chicago: University of Chicago Press, 1980.

Lakoff, George, and Mark Johnson. *Philosophy in the Flesh: The Embodied Mind and Its Challenge to Western Thought.* New York: Basic Books, 1999.

Langer, Suzanne K. *Mind: An Essay on Human Feeling,* Vol. I. Baltimore: Johns Hopkins Press, 1967.

Langer, Suzanne K. *Mind: An Essay on Human Feeling,* Vol. II. Baltimore: Johns Hopkins Press, 1972.

Langer, Suzanne K. *Philosophy in a New Key: A Study in the Symbolism of Reason, Rite, and Art,* 3d Ed. Cambridge, MA: Harvard University Press, 1979.

Langley, L. L. *Homeostasis.* New York: Reinhold, 1965.

Lawler, Robert. *Sacred Geometry: Philosophy and Practice.* New York: Crossroad, 1982.

LeDoux, Joseph. *The Emotional Brain: The Mysterious Underpinnings of Emotional Life.* New York: Simon and Schuster, 1996.

Leyton, Michael. *Symmetry, Causality, Mind.* Cambridge, MA: MIT Press, 1992

Lindsay, Peter H., and Donald A. Norman. *Human Information Processing: An Introduction to Psychology.* New York: Academic Press, 1972.

Longinotti-Buitoni, Gian Luigi (with Kip Longinotti-Buitoni). *Selling Dreams: How to Make Any Product Irresistible.* New York: Simon and Schuster, 1999.

Lord, E. A., and C. B. Wilson. *The Mathematical Description of Shape and Form.* New York: Halstead Press, 1986.

Louch, A. R. *Explanation and Human Action.* Berkeley: University of California Press, 1969.

Lynn, R. *Attention, Arousal, and the Orientation Reaction.* London: Pergamon Press, 1966.

Mandler, George. *Mind and Emotion.* New York: Wiley, 1975.

Margolin, Victor (ed.). *Design Discourse: History, Theory, Criticism.* Chicago: University of Chicago Press, 1989.

Martindale, Colin. *Romantic Progression: The Psychology of Literary History.* New York: Wiley, 1975.

Martindale, Colin. *The Clockwork Muse: The Predictability of Artistic Change.* New York: Basic Books, 1990.

May, Rollo. *Love and Will.* New York: Dell, 1969.

May, Rollo. *The Courage to Create.* New York: Bantam Books, 1973.

McClelland, D. C., et al. *The Achievement Motive.* New York: Appleton-Century-Crofts, 1953.

McIlhany, Sterling. *Art as Design: Design as Art.* New York: Van Nostrand Reinhold, 1970.

Meyer, Leonard B. *Emotion and Meaning in Music.* Chicago: University of Chicago Press, 1956.

Miller, Douglas T., and Marion Nowak. *The Fifties: The Way We Really Were.* Garden City, NY: Doubleday, 1977.

Miller, George A. "The Magical Number Seven, Plus or Minus Two: Some Limits on Our Capacity for Processing Information." *Psychological Review,* Vol. 63, 1956, pp. 81–96.

Mintzberg, H. "The Design School: Reconsidering the Basic Premises of Strategic Management." *Strategic Management Journal,* Vol. 11, 1990, pp. 171–195.

Mitchell, Arnold. *Nine American Lifestyles.* New York: Macmillan, 1983.

Moles, Abraham (translated by Joel E. Cohen). *Information Theory and Esthetic Perception.* Urbana: University of Illinois Press, 1966.

Montagu, Ashley. *Touching: The Human Significance of the Skin.* New York: Columbia University Press, 1971.

Moorhouse, H. F. *Driving Ambitions: A Social Analysis of the American Hot Rod Enthusiasm.* Manchester, England: Manchester University Press, 1991.

Moreland, R. L., and R. B. Zajonc. "A Strong Test of Exposure Effects." *Journal of Experimental Social Psychology,* Vol. 12, 1976, pp. 170–179.

Moreland, R. L., and R. B. Zajonc. "Is Stimulus Recognition a Necessary Condition for the Occurrence of Exposure Effects?" *Journal of Personality and Social Psychology,* Vol. 35, 1977, pp. 191–199.

Morgan, G. *Images of Organization.* Newbury Park, CA: Sage, 1986.

Needham, Rodney (ed.). *Right and Left: Essays on Dual Symbolic Classification.* Chicago: University of Chicago Press, 1973.

Norman, Donald A. *The Design of Everyday Things.* New York: Doubleday, 1990.

Norman, Donald A., and Stephen W. Draper. *User-Centered System Design.* Hillsdale, NJ: Lawrence Erlbaum Associates, 1986

Olds, J., and M. Olds. *Drives, Rewards and the Brain: New Directions in Psychology,* Vol. 2. New York: Holt, Rinehart and Winston, 1965.

Ornstein, Robert. *Evolution of Consciousness: Of Darwin, Freud, and Cranial Fire—The Origins of the Way We Think.* Englewood Cliffs, NJ: Prentice Hall, 1991.

Ornstein, Robert. *The Right Mind: Making Sense of the Hemispheres.* New York: Harcourt Brace, 1997.

Osgood, C. E., et al. *The Measurement of Meaning.* Urbana, IL: University of Illinois Press, 1957.

Osgood, C. E., et al. *Cross-Cultural Universals of Affective Meaning.* Urbana, IL: University of Illinois Press, 1975.

Pacey, Arnold. *Meaning in Technology.* Cambridge, MA: MIT Press, 1999.

Peckham, Morse. *Man's Rage for Chaos: Biology, Behavior and the Arts.* New York: Schocken Books, 1967.

Pedoe, Daniel. *Geometry and the Visual Arts.* New York: Dover, 1976.

Penrose, Roger. *Shadows of the Mind: A Search for the Missing Science of Consciousness.* Oxford, England: Oxford University Press, 1994.

Peters, Tom. *The Circle of Innovation.* New York: Knopf, 1997.

Peters, Tom, and R. H. Waterman, Jr. *In Search of Excellence.* New York: Harper and Row, 1982.

Petroski, Henry. *The Evolution of Useful Things.* New York: Knopf, 1993.

Pevsner, Nikolaus. *Pioneers of Modern Design: From William Morris to Walter Gropius.* New York: Museum of Modern Art, 1949.

Pierce, J. R. *Symbols, Signals and Noise: The Nature and Process of Communication.* New York: Harper and Row, 1961.

Pinker, Steven. *How the Mind Works.* New York: Norton, 1997.

Pirsig, Robert M. *Zen and the Art of Motorcycle Maintenance.* New York: William Morrow, 1974.

Poerksen, Uwe (translated by Jutta Mason and David Cayley). *Plastic Words: The Tyranny of a Modular Language.* University Park, PA: Pennsylvania State University Press, 1995.

Pondy, L. R. "Leadership Is a Language Game." In M. W. McCall, Jr. and M. M. Lombardo (eds.), *Leadership: Where Else Can We Go?* Durham, NC: Duke University Press, 1978, pp. 87–99.

Popcorn, Faith. *The Popcorn Report: On the Future of Your Company, Your World, Your Life.* New York: Doubleday Currency, 1991.

Pribram, Karl H. *Languages of the Brain.* Englewood Cliffs, NJ: Prentice-Hall, 1971.

Pulos, Arthur J. *American Design Ethic: A History of Industrial Design.* Cambridge MA: MIT Press, 1983.

Pulos, Arthur J. *The American Design Adventure.* Cambridge, MA: MIT Press, 1988.

Putnam, Hilary. *Representation and Reality.* Cambridge, MA: MIT Press, 1988.

Resnikoff, Howard L. *The Illusion of Reality.* New York: Springer-Verlag, 1989.

Rifkin, Jeremy. *The Age of Access: The New Culture of Hypercapitalism Where All of Life Is a Paid-for Experience.* New York: Penguin-Putnam, 2000.

Rogers, Everett M. *Diffusion of Innovations,* 3d Ed. New York: Macmillan, 1983.

Runyon, Kenneth E., and David W. Stewart. *Consumer Behavior.* Columbus, OH: Merrill, 1987.

Ryan, Cornelius (ed.). *Conquest of the Moon.* New York: Viking Press, 1953.

Salem, Lionel, et al. *The Most Beautiful Mathematical Formulas.* New York: Wiley, 1992.

Santayana, George. *The Sense of Beauty: Being the Outlines of Aesthetic Theory.* New York: Random House, 1955.

Schultz, D. P. *Sensory Restriction: Effects on Behavior.* New York: Academic Press, 1965.

Scott, Geoffrey. *The Architecture of Humanism: A Study of the History of Taste.* New York: Doubleday Anchor, 1956.

Shannon, C. E., and W. Weaver. *The Mathematical Theory of Communication.* Urbana: University of Illinois Press, 1949.

Simon, Herbert A. *The Sciences of the Artificial,* 2d Ed. Cambridge, MA: MIT Press, 1981.

Singh, Jagjit. *Great Ideas in Information Theory, Language and Cybernetics.* New York: Dover, 1966.

Sluckin, W., et al. "Novelty and Human Aesthetic Preferences." In J. Archer and L. Birke (eds.). *Exploration in Animals and Humans.* London: Van Nostrand, 1983.

Sommer, Robert. *Design Awareness.* San Francisco: Rinehart Press, 1972.

Sparke, Penny. *Design in Context.* Secaucus, NJ: Chartwell Books, 1987.

Stiny, George, and James Gips. *Algorithmic Aesthetics: Computer Models for Criticism and Design in the Arts.* Berkeley: University of California Press, 1978.

Styles, Elizabeth A. *The Psychology of Attention.* Hove, England: Psychology Press, 1997.

Tarnas, Richard. *The Passion of the Western Mind: Understanding the Ideas That Have Shaped Our World View.* New York: Bellantine Books, 1991.

Thackara, John. *Design after Modernism: Beyond the Object.* New York: Thames and Hudson, 1988.

Wechsler, Judith (ed.). *On Aesthetics in Science.* Cambridge, MA: MIT Press, 1978.

Weitz, Morris. *Problems in Aesthetics: An Introductory Book of Readings.* New York: Macmillan, 1959.

Whitney, John. *Digital Harmony: On the Complementarity of Music and Visual Art.* New York: McGraw-Hill, 1980.

Wilentz, Joan Steen. *The Senses of Man.* New York: Thomas Y. Crowell, 1968.

Wilson, Edward O. *Consilience: The Unity of Knowledge.* New York: Knopf, 1998.

Wilson, Elizabeth. *Adorned in Dreams: Fashion and Modernity.* Berkeley: University of California Press, 1985.

Wolf, Michael J. *The Entertainment Economy: How Mega-Media Forces Are Transforming Our Lives.* New York: Random House, 1999.

Wundt, W. M. *Gundzüge der physiologischen Psychologie.* Leipzig: Engelmann, 1874.

Zajonc, R. B., et al. "Effect of Extreme Exposure Frequencies on Different Affective Ratings of Stimuli." *Perceptual and Motor Skills,* Vol. 38, 1974, pp. 667–678.

Zubek, J. P. (ed.). *Sensory Deprivation: Fifteen Years of Research.* New York: Appleton-Century-Crofts, 1969.

CREDITS

p. 3: Timex watches. (*Timex Corporation.*); **p. 7:** Palm V handheld computer. (*Design: IDEO. Photo: Steve Moeder/IDEO.*); **p. 6:** Herman Miller Aeron chair. (*Design: Don Chadwick & Bill Stumpf. Photo: Design Within Reach.*); **p. 6:** 1983 Chrysler Minivan. (*DaimlerChrysler.*); **p. 7:** 1960 Chrysler Imperial. (*Ludvigsen Library LTD.*); **p. 11:** 2001 Chrysler PT Cruiser. (*DaimlerChrysler.*); **p. 18:** Empathic sketch (top) and technical illustration. (*Audi of America, Inc.* [*technical drawing.*]); **p. 19:** Audi TT Coupe and Roadster. (*Audi of America, Inc.*); **p. 28:** Prototype watch designed by Nathan George Horwitt (1947). (*Brooklyn Museum of Art. Gift of Nathan George Horwitt.*); **p. 38:** OXO Good Grips kitchen tools. (*Smart Design.*); **p. 53:** 1936 Cord. (*Ludvigsen Library LTD.*); **p. 58:** Dodge Stratus R/T Coupe (top) and Dodge Neon. (*Daimler-Chrysler.*); **p. 61:** Volkswagen New Beetle. (*Volkswagen of America, Inc.*); **p. 64:** Watches A&B w/Ideal. (*Watch photos: Timex Corporation.*); **p. 74:** Douglas DC-3. (*George H. Stewart.*); **p. 75:** 1934 Chrysler Airflow and Union Pacific Streamliner. (*DaimlerChrysler.*); **p. 76:** 1957 DeSoto. (*Alden C. Jewell.*); **p. 80:** 1957 Cadillac. (*Alden C. Jewell and General Motors Media Archives.*); **p. 81:** Pontiac Rev concept car. (*General Motors Corporation.*); **p. 84:** 1936 Cord. (*Ludvigsen Library LTD.*); **p. 85:** 1941 Graham Hollywood. (*Margaret Bawden.*); **p. 87:** Arm chair designed by Ludwig Mies van der Rohe (1927). (*Design Within Reach.*); **p. 93:** The Mercury locomotive designed for the New York Central Railroad by Henry Dreyfuss. (*Harold K. Vollrath.*); **p. 94:** Lockheed Constellation. (*George H. Stewart.*); **p. 95:** Bent plywood chair designed by Charles Eames (1946). (*Design Within Reach.*); **p. 95:** Cesca chair designed by Marcel Breuer (1928). (*Design Within Reach.*); **p. 114:** Collar MKC43 by Arline M. Fisch. (*The Liliane and David M. Stewart Collection/Montreal Museum of Decorative Arts. Photo: William Gullette.*); **p. 115:** Chrysler 300 concept car. (*Daimler/Chrysler.*); **p. 145:** Corning Corelle cup compared with computer-generated stereotypical cup. (*Computer-generated stereotype: Eric Chen.*); **p. 150:** 2000 and 2001 Chrysler minivans. (*DaimlerChrysler.*); **p. 152:** Movado Museum Watch. (*Movado/NAWC.*); **p. 153:** 2001 Toyota Prius. (*Toyota Motor Sales/Dewhurst.*); **p. 153:** 1995 Mercury Sable. (*Ford Motor Company.*); **p. 158:** 9 Tumblers (*Agota M. Jonas, San Jose State University.*); **p. 209:** Saber saw design. (*T. C. Chang, San Jose State University.*); **p. 212:** Iron design. (*Christopher B. Schmidt, San Jose State University.*); **p. 227:** Empathic sketch and iron design. (*Maz Kattuah, San Jose State University.*)

INDEX

D

E

M

N

ABOUT THE AUTHOR

Del Coates has had a long and distinguished career as a practitioner and professor of industrial design. As a consultant, he has advised an international list of clients. He was a designer with Ford Motor Company's Advanced Vehicle Concepts Department. He was a member of the team that designed Herman Miller's Action Office, the furniture system that revolutionized office landscapes. As director of design research at GVO, a pioneer product design and development firm in Silicon Valley, he contributed to the design of only the second computer to be featured in the design collection of New York's Museum of Modern Art. As design strategist for Nissan Motor Company, he coined the name and concept that became the Maxima.

Del Coates is one of the world's most influential industrial design teachers. Alumni of San Jose State University, where he is a professor of industrial design and ergonomics, have taken leading roles in shaping the products of Silicon Valley—arguably the crucible of modern industrial design—for nearly two decades. Applying principles described in this book, former and current students routinely win top design honors, including those awarded annually by *BusinessWeek* and the Industrial Designers Society of America (IDSA). He also teaches his principles and methods each summer to business, engineering, and design professionals at the University of Michigan's Human Factors Engineering Short Courses.